Sustainable Agriculture

Sustainable Agriculture

Logen Robinson

R CALLISTO
REFERENCE

www.callistoreference.com

Callisto Reference,
118-35 Queens Blvd., Suite 400,
Forest Hills, NY 11375, USA

Visit us on the World Wide Web at:
www.callistoreference.com

ISBN: 978-1-64116-517-4 (Hardback)

Cataloging-in-Publication Data

Sustainable agriculture / Logen Robinson.
    p. cm.
Includes bibliographical references and index.
ISBN 978-1-64116-517-4
1. Sustainable agriculture. 2. Alternative agriculture. 3. Agriculture. I. Robinson, Logen.
S494.5.S86 S87 2022
630--dc23

Table of Contents

Preface

Farming in sustainable ways is referred to as sustainable agriculture. Its key feature is that it seeks to meet the current textile and food needs of the society without compromising the ability of the coming generations to meet their needs. It is based on the understanding of various benefits which can be obtained from well-functioning eco-systems. It can be used to produce adequate amount of food to feed a growing population without affecting the environment in a negative manner. There are a number of practices which are associated with sustainable agriculture such as minimizing the use of water as well as the level of pollution on the farm. This textbook elucidates new techniques and their applications in a multidisciplinary approach. It aims to shed light on some of the unexplored aspects of sustainable agriculture. Those in search of information to further their knowledge will be greatly assisted by this book.

A detailed account of the significant topics covered in this book is provided below:

Chapter 1- The type of farming which makes use of sustainable methods is known as sustainable agriculture. It is based upon the understanding of ecosystem services, as well as the relationships between organisms and their environment. The topics elaborated in this chapter will help in gaining a better perspective about sustainable agriculture and its importance.

Chapter 2- The fundamental tenets of sustainable farming and gardening are economic profit, social responsibility and environmental stewardship. There are numerous forms of sustainable farming methods such as organic farming, biodynamic agriculture and natural farming. This chapter closely examines these types of sustainable farming to provide an extensive understanding of the subject.

Chapter 3- There are a number of different approaches towards sustainable agriculture. A few of them are hydroculture, regenerative agriculture, permaculture and companion planting. The topics elaborated in this chapter will help in gaining a better perspective about these branches of sustainable agriculture.

Chapter 4- There are a number of practices which can make agriculture more sustainable by reducing long-term damage to the soil. A few of them are crop rotation, planting cover crops, drip irrigation, green manuring and multiple cropping. This chapter discusses in detail these agricultural practices.

Chapter 5- The practice of managing the forest resources in such a way that it preserves the health of the forest while fulfilling the requirements of the society is known as sustainable forest management. Some of the practices which are involved in sustainable regulation of forests are coppicing, pollarding and forest gardening. The diverse aspects of sustainable forest management have been thoroughly discussed in this chapter.

It gives me an immense pleasure to thank our entire team for their efforts. Finally in the end, I would like to thank my family and colleagues who have been a great source of inspiration and support.

Finally, I would like to thank the entire team involved in the inception of this book for their valuable time and contribution. This book would not have been possible without their efforts. I would also like to thank my friends and family for their constant support.

Logen Robinson

Chapter 1

Sustainable Agriculture: An Introduction

The type of farming which makes use of sustainable methods is known as sustainable agriculture. It is based upon the understanding of ecosystem services, as well as the relationships between organisms and their environment. The topics elaborated in this chapter will help in gaining a better perspective about sustainable agriculture and its importance.

Sustainable agriculture is a type of agriculture that focuses on producing long-term crops and livestock while having minimal effects on the environment. This type of agriculture tries to find a good balance between the need for food production and the preservation of the ecological system within the environment. In addition to producing food, there are several overall goals associated with sustainable agriculture, including conserving water, reducing the use of fertilizers and pesticides, and promoting biodiversity in crops grown and the ecosystem. Sustainable agriculture also focuses on maintaining economic stability of farms and helping farmers improve their techniques and quality of life.

There are many farming strategies that are used that help make agriculture more sustainable. Some of the most common techniques include growing plants that can create their own nutrients to reduce the use of fertilizers and rotating crops in fields, which minimizes pesticide use because the crops are changing frequently. Another common technique is mixing crops, which reduces the risk of a disease destroying a whole crop and decreases the need for pesticides and herbicides. Sustainable farmers also utilize water management systems, such as drip irrigation, that waste less water.

Importance of Sustainable Agriculture

Sustainable Agriculture is crucial for our future: because of the following reasons:

1. Sustainable agriculture restores and nourishes the soil

Healthy soil leads to healthier plants and animals, resulting in much more nutritious food for people. Healthy soil holds in moisture much more efficiently than depleted soil does, and leads to resilient healthy plants that are not as susceptible to attacks from diseases and pests.

In contrast, the conventional agriculture style of heavy tillage and ever-increasing toxic chemicals that dominates our food system today is very destructive to soil ecology.

Such as a system generally fails to nourish the soil beyond the three primary nutrients of nitrogen, potassium, and phosphorous. Such a simplified mechanistic view of plant-based nutrition and wellness leaves out many different important nutrients, minerals, and healthful plant compounds that plants need, resulting in depleted soil that grows much less nutritious food and leads to nutrient deficiencies in the human population.

This lack of respect for the soil in today's conventional industrial agriculture system is plagued with problems of soil erosion, crops that are susceptible to attacks from disease and pests (thus requiring more toxic chemicals for "protection"), water pollution, and higher susceptibility to drought.

2. Sustainable agriculture works in harmony with nature and not against it

Sustainable farming learns from what nature has to teach us about how natural productivity truly works and applies those lessons to create systems that are both naturally productive and naturally efficient.

Nature generally works through cooperation and collaboration instead of by domination, everything is recycled in some way, and everything functions well within natural limits. Sustainable agriculture takes future generations into account and is regenerative.

In contrast, conventional industrial agriculture is the embodiment of man working to dominate against nature.

What society is in the midst of discovering right now is that our constant strivings to control everything when we farm is largely failing, and nature will always find a way to outdo us. Although we are very clever, we are still hitting nearly every natural limit that exists, and we absolutely must re-examine our relationship with Mother Nature for our own survival.

3. Sustainable farms saves energy

Sustainable agriculture strives to reduce energy use at all levels. Beyond embracing less energy-intensive forms of agricultural production, we can design farming systems that work smarter, not harder, and systems that will become more productive and efficient with each passing year.

By reducing energy use and eliminating the need for fossil fuels, the greenhouse gas emissions produced by the agricultural sector could be reduced dramatically.

In contrast, conventional industrial agriculture is extremely energy-intensive and is heavily reliant upon fossil fuels for both production and transport many thousands of miles from field to plate. In fact, the conventional industrial agriculture is one of the largest sources of greenhouse gases in the world today.

4. Sustainable farming protects and conserves water

Sustainable agriculture systems employ many methods that conserve water, including the use of mulching, drip irrigation, hugelkultur garden beds, creating swales on contour that help to hold water high on the landscape and recharge underground water resources, planting crops that do not require as much water such as an emphasis on perennial crops with deep roots.

Sustainable agriculture also employs methods that help to protect waterbodies from pollution and works to prevent pollution (in these systems, pollution is considered to be "waste" that is leaving the system), such as through the use of filter strips near water bodies and contour farming.

In contrast, conventional industrial agriculture requires a great deal of water for production, and the crops in such systems are generally vulnerable to drought.

5. Sustainable agriculture values diversity

Sustainable agriculture embraces diverse farming systems that incorporate a variety of crops instead of just a few select few monoculture crops.

Such diversity leads to greater resiliency in the face of drought, diseases, and pests, since a sustainable farm is not as dependent upon a single variety of a crop or just a few primary crops for income.

A diverse sustainable farm also incorporates both plant and animal production together within a cooperative system, and is a healthy place for pollinators, people, and wildlife.

In contrast, conventional industrial agriculture generally relies upon a few primary crops, and chemical fertilizers, pesticides, and herbicides to protect their monocrop plants that are more vulnerable to disease and pests. These agricultural chemicals can be very toxic to people, wildlife, and pollinators.

Crop and animal production are generally separate, with animal waste requiring disposal, and crops require their own fertilizers in these "efficient" conventional agricultural systems.

6. Sustainable agriculture provides resilience in a world of climate change

Sustainable farming methods are the cornerstone of the low input agriculture. bullet style, sustainable agriculture conserves energy and reduces greenhouse gas emissions. Sustainable agriculture conserves water, decreasing vulnerability to drought.

With diverse systems that grow a variety of crops, an emphasis on those plants that naturally require less water, and systems that have healthier plants because of the presence of healthier soils, crops in sustainable systems should have greater resilience than those in conventional systems.

In addition, when sustainable agriculture systems incorporate trees and other perennial plants, along with free-range livestock grazing systems, agriculture can actually become carbon sinks.

7. Local sustainable farms support local communities and economies

Local sustainable farms place a great deal of emphasis on local food production.

By localizing our food system, we reinvest our money in our communities, where it continues to circulate within our local area to provide jobs for our friends and neighbors.

8. True sustainable agriculture is good for people, the planet, and is profitable

Sustainable agriculture is not truly sustainable unless it takes into account how people are affected, the health of our planet, and it should be profitable as well.

Goals of Sustainable Agriculture

The goal of sustainable agriculture is to meet society's food and textile needs in the present without compromising the ability of future generations to meet their own needs. Practitioners of sustainable agriculture seek to integrate three main objectives into their work: a healthy environment, economic profitability, and social and economic equity. Every person involved in the food system—growers, food processors, distributors, retailers, consumers, and waste managers—can play a role in ensuring a sustainable agricultural system.

There are many practices commonly used by people working in sustainable agriculture and sustainable food systems. Growers may use methods to promote soil health, minimize water use, and lower pollution levels on the farm. Consumers and retailers concerned with sustainability can look for "values-based" foods that are grown using methods promoting farmworker wellbeing, that are environmentally friendly, or that strengthen the local economy. And researchers in sustainable agriculture often cross disciplinary lines with their work: combining biology, economics, engineering, chemistry, community development, and many others. However, sustainable agriculture is more than a collection of practices. It is also process of negotiation: a push and pull between the sometimes competing interests of an individual farmer or of people in a community as they work to solve complex problems about how we grow our food and fiber.

Principles of Sustainable Agriculture

The principles of sustainable agriculture are:

Supply Chain

Sustainable Agriculture supports a world where producers and consumers, not corporations, control the food chain.

Food Sovereignty

Sustainable agriculture contributes to rural development and fighting poverty and hunger, by enabling livelihoods in rural communities that are safe, healthy, and economically viable.

Food Production and Consumption

Smarter food production and consumption are possible today without impacts on the environment and health to ensure food safety and fight food waste. We must decrease meat consumption, and minimise the use of land for bioenergy. We must also achieve higher yields where they are needed – through ecological means.

Biodiversity

Ecological Farming promotes nature's diversity during all steps of the supply chain, from the seed to the plate through different actions, from seed production to consumption education.

Soil Fertility

Ecological Farming protects and increases soil fertility, by promoting suitable farming practices and eliminating those that consume and contaminate the soil.

Ecological Pest Management

Ecological Farming enables farmers to control pests and weeds – without the use of expensive chemical pesticides that can harm our soil, water and ecosystems, and the health of farmers and consumers.

Strengthen Agriculture

Ecological Farming strengthens our agriculture, and effectively adapts our food system to changing climatic conditions and economic realities.

Methods of Sustainable Agriculture

The various methods of sustainable agriculture are:

- Crop Rotation: Crop rotation is one of the most powerful techniques of sustainable agriculture. Its purpose is to avoid the consequences that come with planting the same crops in the same soil for years in a row. It helps tackle pest problems, as many pests prefer specific crops. If the pests have a steady food supply they can greatly increase their population size. Rotation breaks the reproduction cycles of pests. During rotation, farmers can plant certain crops, which replenish plant nutrients. These crops reduce the need for chemical fertilizers.

- Cover Crops: Many farmers choose to have crops planted in a field at all times and never leave it barren, this can cause unintended consequences. By planting cover crops, such as clover or oats, the farmer can achieve his goals of preventing soil erosion, suppressing the growth of weeds, and enhancing the quality of the soil. The use of cover crops also reduces the need for chemicals such as fertilizers.

- Soil Enrichment: Soil is a central component of agricultural ecosystems. Healthy soil is full of life, which can often be killed by the overuse of pesticides. Good soils can increase yields as well as creating more robust crops. It is possible to maintain and enhance the quality of soil in many ways. Some examples include leaving crop residue in the field after a harvest, and the use of composted plant material or animal manure.

- Natural Pest Predators: In order to maintain effective control over pests, it is important to view the farm as an ecosystem as opposed to a factory. For example, many birds and other animals are in fact natural predators of agricultural pests. Managing your farm so that it can harbor populations of these pest predators is an effective as well as a sophisticated technique. The use of chemical pesticides can result in the indiscriminate killing of pest predators.

- Bio intensive Integrated Pest Management: Integrated pest management (IPM). This is an approach, which really relies on biological as opposed to chemical methods. IMP also emphasizes the importance of crop rotation to combat pest management. Once a pest problem is identified, IPM will mean that chemical solutions will only be used as a last resort. Instead the appropriate responses would be the use of sterile males, and biocontrol agents such as ladybirds.

Benefits of Sustainable Agriculture

There are many benefits of sustainable agriculture, and overall, they can be divided into human health benefits and environmental benefits. In terms of human health, crops grown through sustainable agriculture are better for people. Due to the lack of chemical pesticides and fertilizers, people are not being exposed to or consuming synthetic materials. This limits the risk of people becoming ill from exposure to these chemicals. In addition, the crops produced through sustainable agriculture can also be more nutritious because the overall crops are healthier and more natural.

Sustainable agriculture has also had positive impacts of the environment. One major benefit to the environment is that sustainable agriculture uses 30% less energy per unit of crop yield in comparison to industrialized agriculture. This reduced reliance on fossil fuels results in the release of less chemicals and pollution into the environment. Sustainable agriculture also benefits the environment by maintaining soil quality, reducing soil degradation and erosion, and saving water. In addition to these benefits, sustainable agriculture also increases biodiversity of the area by providing a variety of organisms with healthy and natural environments to live in.

Sustainable agriculture is an approach of farming that includes a wide range of methods of ranching and farming, which result in the benefits for farmers, their families, the environment, and farm animals. The methods involved in sustainable agriculture produces food, which is healthy for users, causes no harm to the environment, humanitarian for workers, and treats animals with respect. It also gives financial benefits to the farmers, and boosts up the rural communities.

Contributes to Environmental Conservation

The environment plays a major role in fulfilling our basic necessities of life. It is therefore out duty to return some of these things back so that our future generations may not remain deprived. Sustainable farming helps in putting back some of these things back to the environment. This helps to replenish land and other resources like soil, water, and air to make them sufficiently available for the coming generations.

Prevents Pollution

When sustainable farming is carried the waste so produced remains inside the farma's ecosystem. Thus it cannot in any way cause pollution or buildup in the external environment.

Reduction in Cost

Sustainable agriculture minimizes the use and cost of purchasing fossil fuel and reduces the transportation costs. This helps in reducing the overall cost involved I the process of farming.

Biodiversity

Sustainable agriculture results in biodiversity as the farms produce different kinds of animals and plants. Plants are seasonally rotated about the fields, which results in enriched soil, prevention of diseases and outbreaks of pests.

Beneficial for Animals

Animals are cared for, treated humanely and with respect. All animals living in the farm are facilitated to exhibit their natural behaviors like grazing, pecking or, rooting. This helps them to grow in a natural way.

Economically Beneficial for Farmers

When farmers engage themselves into sustainable agriculture they receive a fair wage for their effort. As a result their dependence on government subsidies is reduced, thereby strengthening the rural communities.

Social Equality

When sustainable agriculture is practiced workers are offered competitive salaries and benefits. They are treated with humanity; provided with safe work environment, food and proper living conditions.

Beneficial for Environment

Sustainable agriculture decreases the use of non-renewable environmental resources and is thus quite beneficial for the environment. This special type of agriculture and farming technique makes utmost use of the environment and that too without causing any harm to it. Products obtained do not contain any inorganic chemicals like insecticides and pesticides. All these factors make sustainable agriculture a preferred choice of farmers all over the world.

The Environment and Natural Resources

By completely ignoring the impact farming has on natural resources, ancient civilisations have met untimely ends in time gone by. These civilisations carry a warning for the contemporary agriculture industry. Control the application of natural resources and monitor the effects that its techniques have on the environment.

Sustainable farms are able to produce crops and livestock without using copious amounts of synthetic chemicals or GMO's. They will also avoid farming practices that degrade the earth's topsoil or that require large volumes of water or other natural resources. This initially seems like it would have a huge impact on Australian farmers, where drought has become commonplace and restricting water usage may mean huge losses in crop quality and quantity. But the point of the movement is to find sustainable alternatives, and this is exactly what it does.

Water

The necessity of preserving water reserves is so important that farmers are now taking steps before planting crops to ensure that water can be used sparingly throughout the harvest season. This has come in the form of incentives for farmers buying drought resistant crops, low-volume irrigation systems and occasionally avoiding planting altogether.

Salt and pesticides used in water for crops have also become an issue for the environment. Some farms have been converted from wild habitats or are close enough to these habitats that the run-offs introduce foreign levels of salt and fertiliser into those environments. By managing the chemicals in these products, sustainable agriculture can protect native ecosystems and avoid a diversion of local water resources.

These same chemicals that are effecting natural environments are also affecting human ones. Sustainable agriculture attempts to move away from GMO's and towards natural and non volatile crops.

Air Quality and Energy

No environmental discussion would be complete without mentioning the impact of power sources. Just as in any other industry, the discussion surrounding energy remains ambivalent. While there is a clear need to move away from fossil fuels, there is no energy source currently available that will replace the existing systems without having a significant impact on the agricultural economy.

Machinery innovation is already reaching this goal by improving the air quality and the earth's health with conservative designs. EPA standards imposed on Farming machinery producers made a sink or swim situation. These companies responded by building Eco-considerate engines that don't impact the efficiency of farm and agriculture work.

Soil

A healthy topsoil means a healthy industry. Managing a soil's health is a key factor in making sure a farm has a future in cultivation. Topsoil is also an essential element within agricultural ecosystems and by improving its health, a crop can experience an increase in yields.

Aussie farmers are already testing methods to enrich the topsoil for crops. Simple techniques like crop rotation and adding animal manure or decomposing plant matter to fields have had positive effects on the topsoil. Other techniques like planting diverse crops to avoid that whole 'don't put all your eggs in one basket' concept, have resulted in more biodiversity on farmlands.

Animal Well-being

Aside from preventing pollution, sparing our natural resources and looking after our nations farmers, sustainable agriculture will also petition for those without a voice. The movement strives to earn livestock better lives. It does this by letting the animals live naturally within their environments. This signals a depart from the current standard for meat, eggs, and dairy products. Most of which come from factory animals living in cramped cages with limited to no access to the outdoors.

The focus on livestock and animal well-being allows farmers to let their animals live as intended. The effects will potentially reduce industry waste and malpractice. It will also give animals the right to move freely, eat and act naturally, while evading the stress of being cramped in a factory.

Issues Associated with Sustainable Agriculture

Although there are many benefits to sustainable agriculture, there are also some issues associated with it. One of the main concerns is that sustainable agriculture does not produce as much food as industrialized agriculture. A study conducted over two decades in Switzerland showed that crop yield was 20% lower on farms that practiced sustainable agriculture in relation to farms that utilized industrialized methods. The lower productions rates of sustainable agriculture raise concerns that this method will not be able to produce enough food to feed the growing population.

References

- What-is-sustainable-agriculture, definition-benefits-and-issues: study.com, Retrieved 2 July, 2019

- Why-do-we-need-sustainable-agriculture: greentumble.com, Retrieved 6 January, 2019

- What-is-sustainable-agriculture: ucdavis.edu, Retrieved 12 August, 2019

- Greenpeace-sustainable-agriculture: lifegate.com, Retrieved 18 February, 2019

- Methods-and-benefits-of-sustainable-agriculture: conserve-energy-future.com, Retrieved 24 January, 2019

- Benefits-sustainable-agriculture: edgeindiaagrotech.com, Retrieved 30 March, 2019

- What-are-the-benefits-of-sustainable-agriculture: machines4u.com.au, Retrieved 3 February, 2019

- Benefits-sustainable-agriculture: edgeindiaagrotech.com, Retrieved 9 April, 2019

Chapter 2

Sustainable Farming and Gardening

The fundamental tenets of sustainable farming and gardening are economic profit, social responsibility and environmental stewardship. There are numerous forms of sustainable farming methods such as organic farming, biodynamic agriculture and natural farming. This chapter closely examines these types of sustainable farming to provide an extensive understanding of the subject.

Sustainable Farming

Sustainable farming, is using farming practices considering the ecological cycles. It is also sensitive towards the microorganisms and their equations with the environment at large. In simpler terms, sustainable farming is farming ecologically by promoting methods and practices that are economically viable, environmentally sound and protect public health. It does not only concentrate on the economic aspect of farming, but also on the use of non-renewable factors in the process thoughtfully and effectively. This contributes to the growth of nutritious and healthy food as well as bring up the standard of living of the farmer.

Our environment, and subsequently our ecology have become an area of concern for us over the last few decades. This has increasingly led us to contemplate, innovate and employ alternate methods or smaller initiatives to save our ecology. One such initiative is sustainable farming. It simply means production of food, plants and animal products using farming techniques that prove to be beneficial for public health and promote economic profitability. It draws and learns from organic farming.

Sustainable farming helps the farmers innovate and employ recycling methods, this apart from the conventional perks of farming. A very good example of recycling in sustainable farming would be the crop waste or animal manure. The same can be transformed into fertilizers that can help enrich the soil. Another method that can be employed is crop rotation. This helps the soil maintain its nutrients and keeps the soil rich and potent. Collection of rainwater via channeling and then its utilization for irrigation is also a good example of sustainable farming practices.

Benefits of Sustainable Farming

1. Environment Preservation,

2. Economic Profitability,

3. Most efficient use of non-renewable resources,

4. Protection of Public Health,

5. Social and Economic Equity.

Sustainable Farming Methods or Practices

Let us see various methods or practices of Sustainable farming in detail:

1. Make use of Renewable Energy Sources: The first and the most important practice is the use of alternate sources of energy. Use of solar, hydro-power or wind-farms is ecology friendly. Farmers can use solar panels to store solar energy and use it for electrical fencing and running of pumps and heaters. Running river water can be source of hydroelectric power and can be used to run various machines on farms. Similarly, farmers can use geothermal heat pumps to dig beneath the earth and can take advantage of earth's heat.

2. Integrated pest management: Integrated pest management a combination pest control techniques for identifying and observing pests in the initial stages. One needs to also realize that not all pests are harmful and therefore it makes more sense to let them co-exist with the crop than spend money eliminating them. Targeted spraying works best when one need to remove specific pests only. This not only help you to spray pest on the selected areas but will also protect wildlife from getting affected.

3. Crop Rotation: Crop rotation is a tried and tested method used since ancient farming practices proven to keep the soil healthy and nutritious. Crop rotation has a logical explanation to it – the crops are picked in a pattern so that the crops planted this season replenishes the nutrients and salts from the soil that were absorbed by the previous crop cycle. For example, row crops are planted after grains in order to balance the used nutrients.

4. Avoid Soil Erosion: Healthy soil is key to a good crop. Age old techniques like

tilling the land, plowing etc. still work wonders. Manure, fertilizers, cover crops etc. also help improve soil quality. Crop rotations prevent the occurrence of diseases in crops, as per studies conducted. Diseases such as crown rot and tan spot can be controlled. Also pests like septoria, phoma, etc. can be eliminated by crop rotation techniques. Since diseases are crop specific, crop rotation can work wonders.

5. Crop Diversity: Farmers can grow varieties of the same crop yielding small but substantial differences among the plants. This eases financial burdening. This process is called crop diversity and its practical use is on a down slide.

6. Natural Pest Eliminators: Bats, birds, insects etc work as natural pest eliminators. Farmers build shelter to keep these eliminators close. Ladybugs, beetles, green lacewing larvae and fly parasites all feed on pests, including aphids, mites and pest flies. These pest eliminators are available in bulk from pest control stores or farming supply shops. Farmers can buy and release them on or around the crops and let them make the farm as their home.

7. Managed Grazing: A periodic shift of the grazing lands for cattle should be maintained. Moving livestock offers them a variety of grazing pastures. This means they will receive various nutrients which is good for them. The excreta of these animals serves as a natural fertilizer for the land. Change of location also prevents soil erosion as the same patch of land is not trampled upon constantly. Also by grazing in time and mowing the weeds can be gotten rid off before they produce more seeds and multiply.

8. Save Transportation Costs: Targeting the sales of the production in the local market saves transportation and packaging hassles. It also eliminates the need of storage space. Therefore when stuff is grown and sold in local markets, it makes a community self sufficient, economically sound, saves energy and doesn't harm the environment in any way.

9. Better Water Management: The first step in water management is selection of the right crops. One must choose the local crops as they are more adaptable to the weather conditions of the region. Crops that do not command too much

water must be chosen for dry areas. Irrigation systems need to be well planned otherwise they lead to other issues like river depletion, dry land and soil degradation. One can also build rainwater harvesting systems to store rainwater and use them in drought prevailing conditions. apart from that municipal waste water can be used for irrigation after recycling.

10. Removal of Weeds Manually: Farmers having small farms can use their hands to remove weeds from crops where machines can't reach or where crops are too fragile. This is quite a labor intensive task and is not suitable for large farms. Apart from this, a farmer also has the option to burn the old crops so that weeds do not produce seeds and destroy rest of the crops. However, that will cause pollution in air and cal also affect the soil quality.

Sustainable energy is not only economical but it also helps in the conservation of our natural resources. Sustainable farming also helps reduce the need for chemicals fertilizers and pesticides. This makes the process more organic and clean.

Natural Farming

Natural Farming is a farming practice that imitate the way of nature. It can be interpreted in many ways and sometimes people misinterpret the notion of Natural Farming since the word "Natural" is used so casually in many places.

Basis of natural farming is to do nothing. When you are doing nothing, that is nature at work. Doing nothing does not necessarily mean that you do no work at all. It is to remove all the human prejudice from farming and leave it to nature. It almost appears Zen meditation to be one with nature. Natural Farming is developed in Japan primarily by Masanobu Fukuoka and Mokichi Okada.

Most conventional farming practices and researches are done to improve farming by adding this and adding that while Mr. Masanobu Fukuoka started removing unnecessary steps in farming practices such as weeding, pesticide, tilling and fertilizer by questioning himself "How can I NOT do this and that?" Basically, he only saws seeds and harvest. Nature takes care of everything.

Do nothing is the beginning, the conclusion and the way of nature. Realizing that we do not know anything. Nature creates everything and people are there to serve nature.

"Natural Farming" (or "Shizen No-ho" in Japanese) comes from his style of farming close to nature itself. Conventional farming practices are built on human centered biased farming and by removing all these biases natural farming is achieved.

The principles of Nature Farming utilizes and adopts crop production to conform to these dynamic and balanced production systems in nature, which are a result of the interactions of sunlight, water, soil, animals, plants, and microorganisms in natural ecosystems.

It is very important to observe nature without being too confident of our knowledge but with a modest, clear and pure state of mind. In addition, growing good crops requires the development of affection to the crops. Only with such love, a farmer can realize the requirements of soil and for crops to grow healthily, and hence carry out necessary management practices. Agricultural production is a vocation and should be an act of seeking a truly balanced health of all kinds of life forms, including humans as well as the soil, the crops, and the livestock.

Products of Nature Farming, which are safe and full of vitality, have saved various people suffering from diseases, such as children having atopic dermatitis and adult cancer patients. Nature Farming can achieve these. It is a farming method that shows the fundamental way of living to human beings.

Climax Ecosystems

In ecology, climax ecosystems are mature ecosystems that have reached a high degree of stability, productivity and diversity. Natural farmers attempt to mimic those virtues, creating a comparable climax ecosystem, and employ advanced techniques such as intercropping, companion planting and integrated pest management.

No-till

Natural farming recognizes soils as a fundamental natural asset. Ancient soils possess physical and chemical attributes that render them capable of generating and supporting life abundance. It can be argued that tilling actually degrades the delicate balance of a climax soil:

1. Tilling may destroy crucial physical characteristics of a soil such as *water suction*, its ability to send moisture upwards, even during dry spells. The effect is due to pressure differences between soil areas. Furthermore, tilling most certainly destroys soil horizons and hence disrupts the established flow of nutrients. A study suggests that reduced tillage preserves the crop residues on the top of the soil, allowing organic matter to be formed more easily and hence increasing the total organic carbon and nitrogen when compared to conventional tillage. The increases in organic carbon and nitrogen increase aerobic, facultative anaerobic and anaerobic bacteria populations.

2. Tilling over-pumps oxygen to local soil residents, such as bacteria and fungi. As a result, the chemistry of the soil changes. Biological decomposition accelerates and the microbiota mass increases at the expense of other organic matter, adversely affecting most plants, including trees and vegetables. For plants to thrive a certain quantity of organic matter (around 5%) must be present in the soil.

3. Tilling uproots all the plants in the area, turning their roots into food for bacteria and fungi. This damages their ability to aerate the soil. Living roots drill

millions of tiny holes in the soil and thus provide oxygen. They also create room for beneficial insects and annelids (the phylum of worms). Some types of roots contribute directly to soil fertility by funding a mutualistic relationship with certain kinds of bacteria (most famously the rhizobium) that can fix nitrogen.

Fukuoka advocated avoiding any change in the natural landscape. This idea differs significantly from some recent permaculture practice that focuses on permaculture design, which may involve the change in landscape. For example, Sepp Holzer, an Austrian permaculture farmer, advocates the creation of terraces on slopes to control soil erosion. Fukuoka avoided the creation of terraces in his farm, even though terraces were common in China and Japan in his time. Instead, he prevented soil erosion by simply growing trees and shrubs on slopes.

Variants of Natural Farming

Ladybirds consume aphids and are considered beneficial
by natural farmers that apply biological control.

Although the term "natural farming" came into common use in the English language during the 1980s with the translation of the book *One Straw Revolution*, the natural farming mindset itself has a long history throughout the world, spanning from historical Native American practices to modern day urban farms.

Some variants, and their particularities include:

Fertility Farming

In 1951, Newman Turner advocated the practice of "fertility farming", a system featuring the use of a cover crop, no tillage, no chemical fertilizers, no pesticides, no weeding and no composting. Although Turner was a commercial farmer and did not practice random seeding of seed balls, his "fertility farming" principles share similarities with Fukuoka's system of natural farming. Turner also advocate a "natural method" of animal husbandry.

Native American

Recent research in the field of traditional ecological knowledge finds that for over one hundred centuries, Native American tribes worked the land in strikingly similar ways to today's natural farmers. Author and researcher M. Kat Anderson writes that "According to contemporary Native Americans, it is only through interaction and relationships with native plants that mutual respect is established."

Nature Farming (Mokichi Okada)

Japanese farmer and philosopher Mokichi Okada, conceived of a "no fertilizer" farming system in the 1930s that predated Fukuoka. Okada used the same Chinese characters as Fukuoka's "natural farming" however, they are translated into English slightly differently, as nature farming. Agriculture researcher Hu-lian Xu claims that "nature farming" is the correct literal translation of the Japanese term.

Rishi Kheti

In India, natural farming of Masanobu Fukuok was called "Rishi Kheti" by practitioners like Partap Aggarwal. The Rishi Kheti use cow products like buttermilk, milk, curd and its waste urine for preparing growth promoters. The Rishi Kheti is regarded as non-violent farming without any usage of chemical fertilizer and pesticides. They obtain high quality natural or organic produce having medicinal values. Today still a small number of farmers in Madhya Pradesh, Punjab, Maharashtra and Andhra Pradesh, Tamil Nadu use this farming method in India.

Zero Budget Farming

Zero Budget Farming is a variation on natural farming developed in, and primarily practiced in southern India.It also called spiritual farming.The method involves mulching, intercropping, and the use of several preparations which include cow dung. These preparations, generated on-site, are central to the practice, and said to promote microbe and earthworm activity in the soil. Indian agriculturist Subhash Palekar has researched and written extensively on this method.

Perennial Agriculture

Perennial agriculture refers to the practce of, the cultivation of crop species that live longer than two years without the need for replanting each year. Perennial agriculture differs from mainstream agriculture in that it involves relatively less tilling and in some cases requires less labour and fewer pesticides, helping to maintain or even improve soil health. Perennial crops used in perennial agriculture are grown worldwide in various climates and are adapted to local environmental stressors.

Environmental Benefits

Compared with annual crops, perennial crops have extensive root systems, making soil particles difficult to dislodge and thereby limiting soil erosion. Erosion is further reduced by the limited amount of tilling needed to maintain the crop. In addition, since perennial crops do not need to be replanted every year, they require less labour than annuals, and perennials tend to grow rapidly in the spring, enabling them to outcompete annual weeds. They also have adapted over time to deal with local insects and diseases and therefore generally require fewer pesticide applications than annuals. Moreover, their large root systems enable them to cope with environmental stressors, such as drought or irregular rainfall, and they are able to sequester carbon more efficiently than annuals. Perennial grains also potentially could be bred to imitate some of the key aspects of threatened grasslands by providing habitat for wildlife and improving or maintaining soil health, while also producing food for human consumption.

Research and Limitations

Research into perennial grains slowed until the 2000s, when scientists, aided by a better understanding of plant genetics, began to reassess the plausibility and potential of perennial grains. With advancements in genetic technology, such as genetic mapping and genetic engineering, commercially viable perennial grains could be grown on a wide scale in the future. Researchers also found ways to expand the applications of perennial species. With genetic mapping, for example, researchers have been working to identify genes that can be controlled to increase biomass production in perennial species such as switchgrass (Panicum virgatum), an important source of biofuel.

The value of perennial grains, however, is challenged particularly by their limited yields, which is a result of the nature by which perennials continually regrow every year. Annual grains are able to divert much of their energy into the production of seeds, whereas perennials must divert energy to their root systems and seed development.

Advantages of Perennial Agriculture

Agriculture can be made far more sustainable by transitioning many annual agricultural systems to perennials. Perennial crops are crops which are alive year-round and are harvested multiple times before dying. Perennial plants are not new to agriculture; plants such as apples and alfalfa are perennials that are already commercially grown and harvested. However, most farmland is devoted to annual agriculture. Cereals, oilseeds, and legumes – all annuals – occupy 69 percent of global croplands. Many of these staple crops can be replaced by perennials by hybridization and other genetic engineering techniques. According to, ten of the thirteen most grown cereals and oilseeds can be hybridized with perennial plants.

Conversion of annual fields into perennial fields offers many biodiversity-friendly benefits. One of these benefits is reduced soil erosion. Annual farming leaves fields fallow in between growing seasons and offers less root mass throughout the growth cycle. This leaves fields vulnerable to wind and water erosion. This erosion destroys topsoil which then pressures microbial and plant populations. Perennial plants develop much greater root mass and protect the soil year-round. Perennial farming can reduce erosion rates by up to 50 percent.

Another benefit of conversion to perennial is reduced chemical runoff. Farming chemicals such as fertilizers and pesticides are not completely absorbed by crops and the excess migrates into waters. Agriculture is responsible for 70 percent of water pollution in the United States. This water pollution is harmful to biodiversity in numerous ways. One of the most significant is the creation of ocean dead zones which cover thousands of square kilometers. Perennials can reduce agricultural chemical runoff because their extensive root systems are more efficient at absorbing chemicals. For example, annual crops have been shown to lose up to 35 times more nitrogen than their perennial counterparts.

Perennial plants also conserve freshwater better than annuals plants. Annual crops lose up to five times more water than perennials. This means that annual fields require more irrigation which threatens fresh water sources and consequently biodiversity in certain ecosystems.

Finally, perennial agriculture uses less fossil fuel than annual agriculture. Annual systems require fields to be tilled and replanted more often than perennial systems. This incurs a higher fuel usage due to farm machinery. For example, perennial corn farming would reduce fuel usage by 300 million USD of diesel fuel in the United States as opposed to annual corn.

The Land Institute show that conversion of significant portions of annual farmland into perennial farmland is an achievable goal. The primary obstacle to the usage of perennial cereals, oilseeds, and legumes are the lack of economically viable plant strains. Advances in genetics and molecular biology in recent years make the development of favorable perennial strains. Research is ongoing to create perennial strains of wheat, sunflowers, sorghum, legumes, maize, rice, and mustard.

Organic Farming

Organic farming is a production system which avoids or largely excludes the use of synthetically compounded fertilizers, pesticides, growth regulators, genetically modified organisms and livestock food additives. To the maximum extent possible organic farming system rely upon crop rotations, use of crop residues, animal manures, legumes, green manures, off farm organic wastes, biofertilizers, mechanical cultivation, mineral bearing rocks and aspects of biological control to maintain soil productivity and tilth to supply plant nutrients and to control insect, weeds and other pests.

Organic methods can increase farm productivity, repair decades of environmental damage and knit small farm families into more sustainable distribution networks leading to improved food security if they organize themselves in production, certification and marketing. During last few years an increasing number of farmers have shown lack of interest in farming and the people who used to cultivate are migrating to other areas. Organic farming is one way to promote either self-sufficiency or food security. Use of massive inputs of chemical fertilizers and toxic pesticides poisons the land and water heavily. The after-effects of this are severe environmental consequences, including loss of topsoil, decrease in soil fertility, surface and ground water contamination and loss of genetic diversity.

Organic farming which is a holistic production management system that promotes and enhances agro-ecosystem health, including biodiversity, biological cycles, and soil biological activity is hence important. Many studies have shown that organic farming methods can produce even higher yields than conventional methods. Significant difference in soil health indicators such as nitrogen mineralization potential and microbial abundance and diversity, which were higher in the organic farms can also be seen. The increased soil health in organic farms also resulted in considerably lower insect and disease incidence. The emphasis on small-scale integrated farming systems has the potential to revitalize rural areas and their economies.

Advantages of Organic Farming

1. It helps to maintain environment health by reducing the level of pollution.

2. It reduces human and animal health hazards by reducing the level of residues in the product.

3. It helps in keeping agricultural production at a sustainable level.

4. It reduces the cost of agricultural production and also improves the soil health.

5. It ensures optimum utilization of natural resources for short-term benefit and helps in conserving them for future generation.

6. It not only saves energy for both animal and machine, but also reduces risk of crop failure.

7. It improves the soil physical properties such as granulation, good tilth, good aeration, easy root penetration and improves water-holding capacity and reduces erosion.

8. It improves the soil's chemical properties such as supply and retention of soil nutrients, reduces nutrient loss into water bodies and environment and promotes favourable chemical reactions.

Nutrient Management in Organic Farming

In organic farming, it is important to constantly work to build a healthy soil that is rich in organic matter and has all the nutrients that the plants need. Several methods viz.

green manuring, addition of manures and biofertilizers etc can be used to build up soil fertility. These organic sources not only add different nutrients to the soil but also help to prevent weeds and increase soil organic matter to feed soil microorganisms. Soil with high organic matter resists soil erosion, holds water better and thus requires less irrigation. Some natural minerals that are needed by the plants to grow and to improve the soil's consistency can also be added. Soil amendments like lime are added to adjust the soil's pH balance. However soil amendment and water should contain minimum heavy metals. Most of the organic fertilizers used are recycled by-products from other industries that would otherwise go to waste. Farmers also make compost from animal manures and mushroom compost. Before compost can be applied to the fields, it is heated and aged for at least two months, reaching and maintaining an internal temperature of 130°-140 °F to kill unwanted bacteria and weed seeds. A number of organic fertilizers/amendments and bacterial and fungal biofertilizers can be used in organic farming depending upon availability and their suitability to crop. Different available organic inputs are described below.

Organic Manures

Commonly available and applied farm yard manure (FYM) and vermicompost etc. are generally low in nutrient content, so high application rates are needed to meet crop nutrient requirements. However, in many developing countries, the availability of organic manures is not sufficient for crop requirements; partly due to its extensive use of cattle dung in energy production. Green manuring with Sesbania, cowpea, green gram etc. are quiet effective to improve the organic matter content of soil. However, use of green manuring has declined in last few decades due to intensive cropping and socioeconomic reasons. Considering these constraints International Federation of Organic Agriculture Movement (IFOAM) and Codex Alimentarius have approved the use of some inorganic sources of plant nutrients like rock phosphate, basic slag, rock potash etc. in organic farming systems. These substances can supply essential nutrients and may be from plant, animal, microbial or mineral origin and may undergo physical, enzymatic or microbial processes and their use does not result in unacceptable effects on produce and the environment including soil organisms.

Bacterial and Fungal Biofertilizers

Contribution of biological fixation of nitrogen on surface of earth is the highest (67.3%) among all the sources of N fixation. Following bacterial and fungal biofertilizers can be used as a component of organic farming in different crops.

- Rhizobium: The effectiveness of symbiotic N_2 fixing bacteria viz. Rhizobia for legume crops eg. Rhizobium, Bradyrhizobium, Sinorhizobium, Azorhizobium, and Mesorhizobium etc have been well recognized. These bacteria infecting legumes have a global distribution. These rhizobia have a N_2-fixing capability up to 450 kg N ha^{-1} depending on host- plant species and bacterial strains.

Carrier based inoculants can be coated on seeds for the introduction of bacterial strains into soil.

• Azotobacter: N_2 fixing free-living bacteria can fix atmospheric nitrogen in cereal crops without any symbiosis. Such free living bacterias are: Azotobacter sp. for different cereal crops; Acetobacter diazotrophicus and Herbaspirillum spp. for sugarcane, sorghum and maize crop. Beside fixing nitrogen, they also increase germination and vigour in young plants leading to an improved crop stand. They can fix 15-20 kg/ha nitrogen per year. Azotobacter sp. also has ability to produce anti fungal compounds against many plant pathogens. Azotobacter can biologically control the nematode diseases of plants also.

• Azospirillum: The genus Azospirillum colonizes in a variety of annual and perennial plants. Studies indicate that Azospirillum can increase the growth of crops like sunflower, carrot, oak, sugarbeet, tomato, pepper, cotton, wheat and rice. The crop yield can increase from 5-30%. Inoculum of Azotobacter and Azospirillum can be produced and applied as in peat formulation through seed coating. The peat formulation can also be directly utilized in field applications.

• Plant growth promoting rhizobacteria: Various bacteria that promote plant growth are collectively called plant growth promoting rhizobacteria (PGPR). PGPR are thought to improve plant growth by colonizing the root system and pre empting the establishment of suppressing deleterious rhizosphere microorganisms on the roots. Large populations of bacteria established in planting material and roots become a partial sink for nutrients in the rhizosphere thus reducing the amount of C and N available to stimulate spores of fungal pathogens or for subsequent colonization of the root. PGPR belong to several genera viz.Actinoplanes, Azotobacter, Bacillus, Pseudomonas, Rhizobium, Bradyrhizobium, Streptomyces, Xanthomonas etc. Bacillus spp. act as biocontrol agent because their endospores are tolerant to heat and desiccation. Seed treatment with B.subtilis is reported to increase yield of carrot by 48%, oats by 33% and groundnut upto 37%.

• Phosphorus-solubilizing bacteria (PSB): Phosphorus is the vital nutrient next to nitrogen for plants and microorganisms. This element is necessary for the nodulation by Rhizobium and even to nitrogen fixers, Azolla and BGA. The phospho microorganism mainly bacteria and fungi make available insoluble phosphorus to the plants. It can increase crop yield up to 200-500 kg/ha and thus 30 to 50 kg Super Phosphate can be saved. Most predominant phosphorus-solubilizing bacteria (PSB) belong to the genera Bacillus and Pseudomonas.

• Mycorrhizal fungi: Root-colonizing mycorrhizal fungi increase tolerance of heavy metal contamination and drought. Mycorrhizal fungi improve soil quality also by having a direct influence on soil aggregation and therefore aeration and water dynamics. An interesting potential of this fungi is its ability to allow plant access to nutrient sources which are generally unavailable to the host plants and thus plants may be able to use insoluble sources of P when inoculated with mycorrhizal fungi but not in the absence of inoculation.

- Blue green algae (BGA): BGA are the pioneer colonizers both in hydrosphere and xerosphere. These organisms have been found to synthesize 0.8×10^{11} tonnes of organic matter, constituting about 40 percent of the total organic matter synthesized annually on this planet. BGA constitute the largest, most diverse and widely distributed group of prokaryotic microscopic organisms that perform oxygenic photosynthesis. These are also known as cyanophyceae and cyanobacteria. These are widely distributed in tropics; and are able to withstand extremes of temperature and drought. BGA has been reported to reduce the pH of soil and improve upon exchangeable calcium and water holding capacity. The recommended method of application of the algal inoculum is broadcasting on standing water about 3 to 4 days after transplantation. After the application of algal inoculum the field should be kept water logged for about a week's time. Establishment of the algal inoculum can be observed within a week of inoculation in the form of floating algal mats, more prominently seen in the afternoon.

- Azolla: A floating water fern 'Azolla' hosts nitrogen fixing BGA Anabaena azollae. Azolla contains 3.4% nitrogen (on dry wt. basis) and add organic matter in soil. This biofertilizer is used for rice cultivation. There are six species of Azolla viz. A. caroliniana, A. nilotica, A. mexicana, A. filiculoides, A. microphylla and A. pinnata. Azolla plant has a floating, branched stem, deeply bilobed leaves and true roots which penetrate the body of water. The leaves are arranged alternately on the stem. Each leaf has a dorsal and ventral lobe. The dorsal fleshy lobe is exposed to air and contains chlorophyll. It grows well in ditches and stagnant water. This fern usually forms a green mat over water. Azolla is readily decomposed to NH_4 which is available to the rice plants. Field trial have shown that rice yields increased by 0.5-2t/ha due to Azolla application.

Weed Management in Organic Farming

In organic farming, chemical herbicides cannot be used. So weeding can be done only manually. Different cultural practices like tillage, flooding, mulching can be used to manage the weeds. Besides, biological (pathogen) method can be used to manage the loss due to weeds. When the ground is fallow, a cover crop can be planted to suppress weeds and build soil quality. Weeds growth can also be limited by using drip irrigation whenever possible, which restricts the distribution of water to the plant line.

Insect Pest Management

In organic farming, the presence of pests is anticipated in advance and accordingly the planting schedules and locations are adjusted as much as possible to avoid serious pest problems. The main strategy to combat harmful pests is to build up a population of beneficial insects, whose larvae feed off the eggs of pests. The key to building a population of beneficial insects is to establish borders (host crops) around fields planted with blends of flowering plants that the beneficial insects particularly like. Then periodically beneficial insects are released into the fields, where the host crops serve as their home base and attract more beneficial insects over time. When faced with a pest outbreak that cannot be handled by beneficial insects, the used of natural or other organically approved insecticides like neem pesticides is done. The two most important criteria for allowed organic pesticides are low toxicity to people and other animals and low persistence in the environment. These criteria are determined by the National Organic Standards.

Diseases Management in Organic Farming

Plant diseases are major constraints for reductions in crop yield and quality in organic and low input production systems. Proper fertility management to crops through balanced supply of macro and micronutrients and adoption of crop rotation have shown to improve the resistance of crops to certain diseases. Thus one of the biggest rewards of organic farming is healthy soil that is alive with beneficial organisms. These healthy microbes, fungi and bacteria keep the harmful bacteria and fungi that cause disease in check.

Limitations and Implications of Organic Farming

There are a few limitations with organic farming such as:

1. Organic manure is not abundantly available and on plant nutrient basis it may be more expensive than chemical fertilizers if organic inputs are purchased.

2. Production in organic farming declines especially during first few years, so the farmer should be given premium prices for organic produce.

3. The guidelines for organic production, processing, transportation and certification etc are beyond the understanding of ordinary farmer.

4. Marketing of organic produce is also not properly streamlined.

Biodynamic Agriculture

Biodynamic agriculture has many similarities with organic agriculture, like abstaining from using synthetically produced plant treatments. But there are telling differences: In viticulture, for example, biodynamic farming limits the use of copper sulphate, a fungicide, to 15kg per hectare over the course of five years, for an average of 3kg copper sulphate per hectare per year. This is half the allowance given in organic farming.

The causes of this limitation are simple. Copper sulphate is a fungicide and, when it is taken into the soil, it has a direct impact on the microflora and mycorrhizae – notably by destroying them. To limit the use of copper sulphate, infusions, teas and decoctions of traditional medicinal plant like stinging nettle, chamomile blossoms and common valerian are sprayed. The natural properties contained in these plants permeate the water of the teas and infusions, offering protection against mildew and allowing the use of a much lower amount of copper sulphate than in organic agriculture.

Biodynamic agriculture aims to produce the best possible in ways that allow future generations to obtain the same – or even better – results. The soil itself is key in this type of agriculture, because it is ultimately dirt on which all life on earth depends, and in particular, the first metre of topsoil.

This humus-rich topsoil has taken thousands of years to get to its present state of frenetic microbiological activity and abundant microflora. These micro-organisms work full-time producing the organic matter that plants need for development. It is clear, then, that we need to take care of them. Conventional farming has destroyed the majority of these microscopic populations with fungicides, insecticides and herbicides to the point that it is now necessary to do an urgent about-face.

Biodynamic Preparations

The biodynamic preparations are produced with natural substances. They are applied in minute doses to enhance soil life, plant growth and quality as well as animal health. There are different kinds of preparations for certain application purposes: Field or spray preparations (horn manure and horn silica), compost preparations (yarrow, chamomile, nettle, oak bark, dandelion and valerian preparation). Furthermore, special preparations such as a horsetail decoction and the ash preparations in order to control weeds and pests.

The preparations are a non-replaceable element of biodynamic agriculture. They are an important help for producing food with Demeter quality. Their use is a compulsory requirement of the Demeter standards.

Principles of how to Produce the Preparations

The production of the preparations takes place on the farm. The preparations are made with certain plant materials, cow manure or quartz meal. The materials are placed in certain animal organs as a cover and fermented in the soil at least half a year. Before using the preparations remaining residues of animal organs are removed.

Even with the production of the preparations it is intended to remain in connnection with biological processes. The function of the animal organs is to concentrate the constructive and formative living forces from the surroundings to the substances within the organs. With this special way of production, being comparable to the potentiation process of homeopathic remedies, the preparations develop a strong yet subtle power.

Application and Mode of Action of the Preparations

The application rate of the biodynamic spray preparations are 300 gram per hectare horn manure and 5 gram per hectare horn silica. The compost preparations are applied with quantities of 1-2 cm^3 each per 10 m$_3$ compost, farmyard manure or liquid manure. The mentioned amounts of horn manure and horn silica are vigorously stirred in 20-50 litres of water per hectare for one hour. As soon as possible after stirring the preparations should be evenly sprayed out on acres and grassland.

The compost preparations are brought spotwise into the organic material. According to Rudolf Steiner, they radiate their forces into the compost. Further specific application methods are described by Wistinghausen et al. The turnover processes in organic fertilizers are stimuleted by the preparations. The intensive stimulation of soil life by prepared fertilizer can be measured by some characteristics, e.g. increase of humus content or enzyme activities or more intensive root growth. Examples of a better product quality caused by the preparations are lower storage losses, reduced nitrate contents and higher contents of sugar and vitamins.

The preparations' mode of action is to stimulate harmonizing living processes. There is no direct nutrient effect of the preparations. The preparations support the self regulation of biological systems.

Sustainable Gardening

Sustainable gardening is a method of growing plants such that the garden is able to successfully sustain itself without requiring many outside resources, pesticides, or herbicides. Sustainable gardening is always organic.

The benefits of sustainable gardening are many, including less reliance on watering and no chemical or pesticide use.

Sustainable gardening works in complete harmony with nature and is good for the environment, because it creates the diversity and balance to not require outside resources such as pesticides, herbicides, or fertilizers.

A sustainable garden is grown using soil amendments such as compost or other organic matter and by using plants that work harmoniously together. Natural rather than chemical methods are used to adjust the pH, soil composition, and provide nutrients.

A completely sustainable garden needs to attract the right predatory or beneficial insects, have enough diversity through companion planting to minimize weed growth, and provide a soil that's healthy enough to retain moisture and is rich in nutrients.

The practices that help in sustainable gardening are:

Conserve Water and Control Water Runoff

- Use drip irrigation or soaker hoses instead of oscillating sprinklers as they result in less water loss due to evaporation.

- Position watering devices to prevent water loss by water falling in storm gutters, walkways or in the street.

- Mulch beds to help retain soil moisture.

- Set up a rain barrel to collect rain water for watering plants.

- Plant a rain garden or develop a swale to help retain water in the soil and prevent runoff.

- Install a cistern to collect water to use for plants, washing clothes, bathing and other non-potable uses as local ordinances allow.

- Investigate the use of grey water use in your area.

- Remove hard surfaces in your landscape to allow water to percolate into the soil and not run off in storm gutters. Replace with a porous surface if needed.

- Incorporate rainscaping features such to manage stormwater.

- Don't use the hose to wash off your driveway, deck or walkway. Instead use a broom or an electric blower. Gas-powered blowers produce more pollutants.

Reduce Fossil-fuel Energy use

- Get some exercise and do some hand digging.

- Pull weeds by hand. This is often more effective and less damaging than resorting to chemical sprays.

- Add landscape lighting only where it is really needed. And when used, use compact fluorescent bulbs or solar-powered lights. Low voltage lighting also uses less electricity and is safer for outdoor use.

- Cut down on holiday lights and invest in the new LED lights that use a lot less energy.

- Demand higher accountability of local governments for their expenditures. Do we really need all the night light pollution around us? As energy prices rise demand that local governments focus on what is most important in their expenditures. Reducing expenditures on lighting buildings, parking lots, gardens, etc at night may just be a waste of money that could be better spent elsewhere.

Deal with Yard and Garden "Waste" in a Sound way

- Develop your own compost pile so you can return the valuable plant material back to the soil in your yard.

- Don't send plant-based garden waste to a landfill. Instead support your local yard waste recycling program for any materials you can't compost and use in your own yard.

- Reuse plastic, clay and other pots in your garden. Don't send them to a landfill. And, when a plastic pot has enjoyed a good life, send it to be recycled. In St. Louis the Missouri Botanical Garden has offered a pot recycling service since 1998.

- If you want to use a chipper-shredder for light use, electric ones result in less air pollution than gas-powered.

Plant Selection

- Replace plants that require a lot of watering with plants that are more drought tolerant. Native plants may be good choices.

- Select plants that perform well in your area and have few problems. In the lower Midwest the Plants of Merit program offers some excellent plants for the area.

- Promote diversity in your yard and garden. Plant a wide variety of plants, which can provide habitats for beneficial insects and reduce damage from periodic diseases. You are also helping to preserve genetic diversity.

- Avoid planting invasive plant species.

Garden Design

- Locate trees to help shade and cool your home in the summer to reduce energy costs. By selecting deciduous trees you can still benefit by receiving warming winter rays.

- Plant a windbreak to reduce winter heating bills.

- On new construction, a green roof might be an option.

- Use only Forest Stewardship Council (FSC) wood for decks, fences, and other garden structures. This certification help guarantee that the wood was produced in a responsible, sustainable way.

- Support movements that preserve corridors of native plants in your area.

- Incorporate rainscaping features such as rain gardens, bioswales and rock dams to manage stormwater.

Plant Maintenance

- Learn to tolerate minor insect damage in your yard and garden and work to increase the number of beneficials. Learn to distinguish the good from the bad! Spraying with a pesticide can place harmful chemicals in the environment and may also kill beneficials or damage nearby plants.

- Learn which plant diseases are harmful to your plants and may warrant control and which are just a cosmetic nuisance that will not affect the health of your tree, shrub, or perennial. For example, leaf spot diseases and leaf galls are very common on trees but few if any require treatment.

- Get a soil test before you add fertilizer and or lime to your yard or garden and follow the recommendation. Over fertilizing can lead to excess plant growth, which can be more susceptible to diseases. Trying to grow a plant in a soil out-side its recommended pH range will result in poor growth or death. Also, fertilizer runoff can pollute streams and groundwater.

Sustainable Landscaping

Sustainable landscaping is the practice of using strategic methods for business and residential landscapes with the purpose of offsetting negative environmental impacts. As each region's local climate and soil quality varies, so do the methods used in sustainable landscaping. Water quality, plants chosen, and products such as fertilizers are selected with the goal of conservation, reducing chemical usage, and preventing erosion. Sustainable landscaping involves multiple areas and focuses on the original plan and design for landscapes as well as the construction stage. Additionally, environmentally friendly methods are used for implementing a sustainable landscape as well as ensuring that it is ecologically maintained.

The field of sustainable landscaping continues to evolve. As the environment faces new challenges, man creates more technologies to address each problem. Whatever the environmental issue at hand, sustainable landscaping aims to correct it. In flood-prone areas, sustainable landscaping may focus on methods used to prevent soil erosion, while in desert areas, the emphasis may be on water conservation. Sustainable landscaping is the practice of assessing a region's environmental and ecological challenges and providing solutions. Sustainable landscaping provides several benefits and advantages. In addition to being environmentally friendly, many methods used save resources, water, time, and energy spent on maintenance. Many sustainable methods are cost-efficient over the long term.

The main goals of sustainable landscape design are to conserve water and energy, reduce waste and decrease runoff. In order to achieve these goals residential gardens should treat water as a resource, value soil, preserve existing plants and conserve material resources.

The principles governing sustainable landscaping are:

Treat Water as a Resource

The demand for water is at an all-time high. Wasteful irrigation accounts for over one-third of the residential water use in the United States. Additionally, rainwater is treated as waste and allowed to flow into gutters and sewers.

A sustainable landscaping approach would be to treat water as a valuable resource. With proper design and plant selection, the need for irrigation can be reduced or eliminated. Furthermore, rainwater harvesting can be to capture stormwater on site and use it for irrigation.

Value your Soil

It's likely that your garden's soil is compacted. Compacted soil leads to problems such as restricted plant growth, erosion, runoff and flooding. Runoff caused by compacted soils is one of the main sources of water pollution.

Preserve Existing Plants

Many homeowners want to remove all the plants from their property so that they can start with a clean slate. Often this ends up doing harm because it disrupts the natural processes occurring in the yard.

A sustainable landscaping approach would be to assess the existing plant material and preserve native plants. Invasive, non-native plants should be removed and replaced with a more appropriate choice. Right plant, right place is a popular saying that should guide your plant selection.

Conserve Material Resources

The typical American landscape produces high amounts of yard and construction waste. Additionally, many of the hardscape materials used are energy-intensive and transported hundreds, or even thousands of miles.

A sustainable landscaping approach would be to reduce yard waste by selecting appropriately sized plants and reusing and recycling construction waste. Furthermore, building materials should be carefully selected, using locally sourced materials whenever possible.

Benefits of Sustainable Landscaping

Besides being environmentally responsible, sustainable landscaping has many benefits for homeowners looking to cut down on the time, effort and resources needed to maintain a lawn and garden.

- Cost-Effective: As their name suggests, sustainable landscapes are self-sustaining. This means they need less tending than traditional landscapes, saving you time and money in the long run.

- Fewer Pesticides: By using native plants and trees with their own defenses against pests, sustainable landscapes don't need harsh pesticides to survive.

- Less Watering: Put down your watering cans. Sustainable landscapes do great with very little extra watering because they're designed with local rain levels in mind.

With the right sustainable landscape design, you can give your lawn and garden the power to thrive on Mother Nature alone.

Ideas for Planning a Sustainable Landscape

Creating a sustainable landscape takes some planning. Before diving in, make sure to read up on your own local climate to find the most efficient landscaping solutions for your environment. Once you've got that information, use these tips to bring sustainable practices into your own backyard.

Choose your Plants Wisely

The plants you choose have to be well-adapted to your local environment. Here are some quick tips for choosing the right foliage for your landscape:

- Get rid of problem plants: Every landscape has them – those plants that just won't thrive. If you have plants that need a lot of nurturing, it might mean they're not right for your climate. Replace them with plants that will do better with your local weather patterns.

- Use Native plants & trees: Native plants often need less water than foreign species, as they're used to the local rain levels. They also have developed strong defenses against local pests and diseases. As an added bonus, they tend to attract helpful neighbors like butterflies and bees who keep your flowers alive and well.

- Beware of Invasive Species: While not all non-native plants are harmful, some can disrupt the natural balance of your landscape. These are called invasive species and they are nature's bullies. They grow so aggressively that they can push out local plants and pollinators, upsetting an entire ecosystem. Be sure to check out the Environmental Protection Agency's What to Plant list before choosing your plant-life.

- Create layers to reduce pests: Layering is essential to creating sustainable landscaping. It mimics natural plant growth and provides a lush environment for nature's best pest control: birds. Invite them to stay by providing various shrubs, trees and other plants for them to call home.

- Pay attention to tree sizes: Regular tree trimming is both costly and often requires power tools which use gas or electricity. Choose trees and shrubs that will fit in at their tallest without extra pruning.

- Consider lawn substitutes: If you're tired of mowing the lawn, ditch grass altogether. Moss, ground cover, or even turf are great no-mow lawn solutions. If you're partial to grass, try a meadow lawn. These need mowing just once or twice a year, and give your landscape a beautiful, natural look. For the best results, be sure to use grass that is native to your local area. Another alternative to traditional lawns is a practice called xeriscaping.

Keep your Soil Healthy

Healthy soil is the foundation of a great, sustainable landscape, as it nourishes your plant-life and prevents erosion.

- Aerate your lawn: Aeration puts small holes in your lawn to allow water and nutrients to get to your grass roots more easily. Aerating once or twice a year will improve your soil quality and keep your grass healthy and strong.

- Start your own compost pile: Composting is an eco-friendly way to keep soil moist and healthy year-round. To give your flowerbeds a sustainable glow, start your own at-home compost pile.

- Set your mower high: Set your mowing deck around 3.5 inches for a more sustainable lawn. According to Purdue University, mowing just the top 1/3 of your grass keeps roots strong, which prevents soil erosion and nutrient loss.

Tackle Rainwater Runoff

Rainwater runoff is a common problem in landscape designs. It can cause erosion and carry pollutants to your community's watershed.

- Use porous surfaces: Concrete walkways can't absorb water and can create mini-rivers that can damage your land. Choose absorbent materials like pea gravel or decomposed granite to reduce your rainwater runoff.

- Keep your yard clean: Whatever you leave in your yard will make its way to your local water system via rainwater runoff. Keep lawn clippings, pet waste and other unsavory materials off your lawn to keep your community's watershed clean.

- Make your own rain barrel: Put that excess rainwater to good use with your very own rain barrel. You'll create a sustainable water supply for your landscape, helping you cut down on your water bill.

- Create a swale: A permaculture swale is a small depression designed to redirect rainwater runoff. These are usually situated at the lowest point in your yard and absorb excess water during downpours. The extra water collected in the swale keeps the rest of your yard moist and healthy in the days following a rainstorm.

Example of a grass-lined swale. Swales can also be lined with rocks to help filtrate rainwater before it reaches the soil.

Hardscape with Eco-Friendly Materials

Another important aspect of sustainable landscaping is using environmentally responsible hardscaping materials.

Solar Lights: Looking to light the way through your new landscape? Opt for solar powered lawn lights to cut down on your electricity use.

Use Reclaimed Materials: One man's trash can be another man's landscaping materials. Visit your local salvage store or browse the 'Free' section on Craigslist.com to find anything from lumber to scrap metal and even used patio furniture. You'll save yourself the cost of purchasing brand new materials and you'll prevent the items from winding up in a landfill.

Avoid PVC: PVC pipes are made using an inefficient, high emissions process and they are not biodegradable. If you're looking for sustainable piping for landscape irrigation or drainage, try recyclable HDPE pipe instead.

HDPE piping is a more sustainable landscaping alternative to PVC.

Maintain Sustainably

No landscape is totally maintenance-free, but here are a few eco-friendly switches you can make to keep yours beautiful and functional.

- Eco-friendly fertilizers: During hot or cold spells, even the most sustainable landscapes can use a little extra love. After a tough summer or winter, use organic fertilizers like bloodmeal, cow manure or compost to help your plants and soil bounce back. It will also encourage earthworms to make their home in your garden, which will help your soil stay aerated naturally.

- Non-toxic pesticides: Though sustainable landscapes are great at handling pests on their own, some bugs don't know how to take a hint. To keep them at bay in a sustainable way, you can make your own non-toxic pesticides right at home.

- Man-Powered Tools: Lawn mowers consume over 1 billion gallons of gasoline in the U.S. every year. Cut down on your gas usage by using manual tools like push reel mowers and manual clippers.

- Go Electric: If your lawn is too big for a manual mower, switch to an electric version. In addition to being emissions-free, electric mowers use only about $5 of electricity annually, so you can save some green by going green.

Xeriscaping

Xeriscaping is the practice of designing landscapes to reduce or eliminate the need for irrigation. This means xeriscaped landscapes need little or no water beyond what the natural climate provides.

Xeriscaping has been embraced in dry regions of the western United States. Prolonged droughts have led water to be regarded as a limited and expensive resource. Denver, Colorado, was one of the first urban areas to support xeriscaping. That citys water department encouraged residents to use less of the city's drinkable water for their lawns and gardens.

Xeriscaping has become widely popular in some areas because of its environmental and financial benefits. The most important environmental aspect of xeriscaping is choosing vegetation that is appropriate for the climate. Vegetation that thrives with little added irrigation is called drought-tolerant vegetation. Xeriscaping often means replacing grassy lawns with soil, rocks, mulch, and drought-tolerant native plant species. Trees such as myrtles and flowers such as daffodils are drought-tolerant plants.

Plants that have especially adapted to arid climates are called xerophytes. In desert areas like Phoenix, Arizona, xeriscaping allows gardeners to plant native xerophytes such as ocotillo.

Supporters of xeriscaping say it can reduce water use by 50 or 75 percent. This saves water and money. In Novato, California, residents were offered conservation incentives (reductions in their water bills) to convert from traditional lawns to xeriscaping. The citys water department estimated that the houses that chose xeriscaping saved 120 gallons of water a day.

Another main component of xeriscaping is installing efficient irrigation methods. Drips and soaker hoses direct water directly to the base of the plant and prevent the water evaporation that sprinklers allow. More efficient irrigation is also achieved when types of plants with similar water needs are grouped together. A xeriscaped landscape needs less maintenance than an area landscaped with grass and water-intensive plants.

Drought-tolerant Plants

The most common example of a xeriscape-friendly plant is the cactus, which has hundreds of different species that are native to North and South America. Cacti have evolved many physical adaptations that conserve water. For example, their prickly spines, the cactus version of leaves, protect the plants from water-seeking animals. Their large, round stems have thickened to store large amounts of water. Their waxy skin reduces water lost to evaporation.

Cacti are far from the only plants appropriate for xeriscaping. Other drought-resistant plants include agave, juniper, and lavender. Many herbs and spices are used in xeriscaping, such as thyme, sage, and oregano. Some plants used for food are drought-resistant, such as black walnuts, Jerusalem artichokes, and sapodilla, a sweet fruit native to Mexico.

Xeriscaping saves water in arid areas.

Xeriscape is a systematic concept for saving water in landscaped areas. Xeriscape is a method of landscaping that promotes water conservation. Rather than a specific "look" or a limited group of plants, Xeriscape is a combination of seven basic landscaping principles. These principles are explained below in the order a landscape planner or property owner would consider to install the best landscape. Each principle must be considered during the planning and design phase, but the sequence of installation is also very important in assuring a successful Xeriscape.

Aquacraft

The seven principles of Xeriscape are:

Planning and Design

Trying to create a landscape without a plan is like trying to build a home without blueprints. A plan provides direction and guidance and will ensure that water-conserving techniques are coordinated and implemented in the landscape. The first step is to look at your existing landscape and create a "base plan." This is a to-scale diagram showing the major elements of your landscape - your house, driveway, sidewalk, deck or patio, existing trees, etc.

Soil Improvements

If you have good topsoil you can ignore this topic. Many people, however, have inferior topsoil because of sand and clay. Clay soil is dense, slow to absorb and release water. If water is applied to clay soil too quickly, it either pools on the surface or runs off. Over watering heavy clay soil can actually drown plants.

On the other hand, sandy soil can't hold water. Unless irrigated frequently, plants in sandy soils tend to dry out.

To enable your soil to better absorb water and allow for deeper roots, you may need to add a soil amendment before you plant. For most soils, adding 1 to 2 inches (2.54 cm to 5.08 cm) of organic matter such as compost or well-aged manure to your soil can be beneficial. Rototill the organic matter into the soil at least 6 inches (15.2 cm) deep.

Efficient Irrigation

A Xeriscape can be irrigated efficiently by hand or with an automatic sprinkler system. If you're installing a sprinkler system or upgrading an existing system, it's a good idea to plan this at the same time you design the landscape. Zone turf areas separately from other plantings and use the irrigation method that waters the plants in each area most efficiently.

Zoning of Plants

Different areas in your yard get different amounts of light, wind, and moisture. To minimize water waste, group together plants with similar light and water requirements, and place them in an area in your yard which matches these requirements. A good rule of thumb is to put high water-use plantings in low lying drainage areas, near downspouts, or in the shade of other plants. It's also helpful to put higher water-use plants where it is easy to water.

Dry, sunny areas or areas far from a hose are great places for the many low water-use plants. Planting a variety of plants with different heights, color and textures creates interest and beauty.

By grouping your plants appropriately, you minimize water waste while ensuring that your plants will flourish in the right environment.

Mulches

Mulching is a great addition to your garden and landscape. Mulch helps keep plants roots cool, prevents soil from crusting, minimizes evaporation, and reduces weed growth. Mulches also give beds a finished look and increase the visual appeal of your garden. Organic mulches, such as bark chips, pole peelings or wood grindings, should be applied at least 4 inches (10.2 cm) deep. Because they decompose over time, they're an excellent choice for new beds. As plants mature and spread, they'll cover the mulched areas.

Inorganic mulches include rocks and gravel, and should be applied at least 2 inches (5.1 cm) deep. They rarely need replacement and are good in windy spots. However, they should not be placed next to the house on the sunny south or west sides, because they tend to retain and radiate heat. Mulch may be applied directly to the soil surface or placed over a landscape fabric.

Turf Alternatives

Traditionally, the landscape of choice across North America has been a carpet of bluegrass turf. Bluegrass is lush and hardy, but in many regions it requires a substantial amount of supplemental watering.

One way to reduce watering requirements is to reduce the amount of bluegrass turf in your landscape. Native or low water use plantings, patios, decks or mulches can beautify your landscape while saving water. Choosing a lower water using turf also serves the same purpose. Such choices can include buffalo grass, blue grama grass, turf type tall fescue and fine fescues.

Appropriate Maintenance

Preserve the beauty of your Xeriscape with regular maintenance. The first year or two, your new landscape will probably require a fair amount of weeding, but as plants mature they will crowd out the weeds, significantly reducing your maintenance time.

In addition to weeding, your Xeriscape will need proper irrigation, pruning, fertilizing and pest control. Maintenance time for a new garden is similar to a traditional landscape, but it decreases over time. In addition to weeding, proper irrigation, pruning, fertilizing and pest control will keep your landscape beautiful and water thrifty.

References

- Sustainable-farming-practices: conserve-energy-future.com, Retrieved 13 August, 2019

- What-is-natural-farming: maunakeatea.com, Retrieved 11 May, 2019

- Perennial-agriculture: britannica.com, Retrieved 21 July, 2019

- Colin Adrien MacKinley Duncan (1996). The Centrality of Agriculture: Between Humankind and the Rest of Nature. McGill-Queen's Press - MQUP. ISBN 978-0-7735-6571-5

- Perennial-agriculture: mit.edu, Retrieved 11 June, 2019

- Organic-farming, crop-production, agriculture: vikaspedia.in, Retrieved 10 January, 2019

- Biodynamic-agriculture, agriculture: dahu.bio, Retrieved 9 May, 2019

- Sustainable-gardening: maximumyield.com, Retrieved 19 February, 2019

- Sustainable-gardening, help-for-the-home-gardener, gardens-gardening: missouribotanicalgarden.org, Retrieved 7 June, 2019

- Sustainable-landscaping-composting-and-going-green-outdoors: tigerturf.com, Retrieved 17 August, 2019

- Sustainable, landscape-design: landscapingnetwork.com, Retrieved 27 March, 2019

- What-is-sustainable-landscaping: dumpsters.com, Retrieved 30 May, 2019

- Xeriscaping: nationalgeographic.org, Retrieved 3 April, 2019

- Xeriscape-Introduction: allianceforwaterefficiency.org, Retrieved 17 July, 2019

Chapter 3

Approaches in Sustainable Agriculture

There are a number of different approaches towards sustainable agriculture. A few of them are hydroculture, regenerative agriculture, permaculture and companion planting. The topics elaborated in this chapter will help in gaining a better perspective about these branches of sustainable agriculture.

Hydroculture

Hydroculture is a method of growing plants in a soilless growing medium (such as clay pebbles) or an aquatic environment. Although plants are often seen growing in soil, it is not essential for plant growth. Instead, plants require 16 chemical elements to thrive, all of which can be delivered without the use of soil. Instead, these chemical elements or "plant nutrients" can be delivered directly in water or through a range of growing media such as rockwool and clay pebbles.

The collective name for soilless gardening is often simply called "hydroculture" and stems from the word "hydro" which is derived from the Greek word meaning water. So "hydroculture" simply means "water culture" to illustrate that plants are grown in water instead of soil.

Basic Hydroculture (Passive Hydroponics)

In basic hydroculture, also known as "passive hydroponics" plants are delivered water and nutrients through capillary action, where the plants absorb the nutrients and water at the rate they need each.

An Example Basic Hydroculture Set-up

There are many variations to the hydroculture method of growing plants but a commonly used, basic hydroculture set-up will consist of an inert growing medium (often expanded clay pellets such as the trademarked LECA and Hydrocorn), culture pots, a water level indicator (which lets you know when to top up your water/nutrient solution), pot liners, and fertiliser specifically tailored for plants grown in hydroculture/hydroponic environments.

Uses and Applications

Hydroculture-based plant cultivation is currently more common in the Netherlands and Europe than other parts of the world. However, it is used the world over as it is a simple-to-use, hygienic and sustainable method of cultivation.

As hydroculture is a form of soilless gardening it can be particularly useful in areas where soil is scarce, such as desert areas or in urban environments where the use of soil is impractical and in some cases impossible. Hydroculture is often used in indoor environments as once established, a hydroculture-based system is usually cheaper and easier to maintain than a soil-based system.

A hydroculture system is often made up of five simple parts – clay pebbles (or a similar inert growing medium), culture pots, water level indicator, pot liners, and fertiliser.

Clay Pebbles

Clay pebbles are the primary growing medium used in hydroculture systems. These expanded clay pellets take the place of soil. They are highly porous which means they are great for growing plants. Clay pebbles are effective at retaining moisture and nutrients, are fully inert, free from soil-borne pests and diseases, provide plenty of oxygenation at the root zone, and give your plants a sturdy support structure to grow and thrive.

It is worth noting that although clay pebbles are the most common growing media used in hydroculture; other inert mediums such as perlite can also be used.

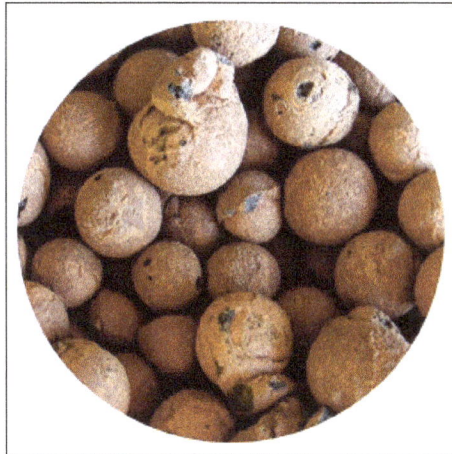

Culture Pots

Culture pots are almost identical to any other plant pot out there on the market, but they have one key difference. A culture pot includes a recess for a water level indicator which will show you exactly how much water your plants are sitting in and giving you a clear indication of when your plants need watering with a nutrient solution.

Water Level Indicator

Water level indicators are a simple way to measure exactly how much water your plants are sitting in. They are tailored for use with specific culture pots and usually come with indicators showing the 'minimum', 'maximum', and 'optimum' water levels. Water level indicators take the guess work out of knowing when to feed your plants.

Pot Liners

Pot liners are often used in hydroculture and other forms of gardening to make easy work of moving plants between containers. They also make porous pots used for display purposes in container gardening waterproof and enourage the development of a smaller, more manageable root zone in plants placed in large outer containers.

Fertilizer

Unlike growing plants in soil, the growing medium used in hydroculture is inert and lacking any plant nutrition. Therefore, plant fertilisation products have been developed specifically for hydroculture and hydroponic applications. There are various fertilisers developed for use during different stages of the growth cycle and they all contain the essential and beneficial elements and nutrients plants need for healthy, vigorous growth.

Advantages of Hydroculture

The following are the main benefits of hydroculture:

1. No fungus gnats: Interior plants are notorious breeding grounds for small flies called "fungus gnats." Although not harmful to the plant, fungus gnats are incredibly annoying to people and quite difficult to control. They thrive in moist or damp organic matter and more typical soils provide this environment, which is why fungus gnats are so common on soil-based plants. LECA is an inorganic (rock-based) growing medium and fungus gnats cannot reproduce or thrive in this environment, thus are non-existent with hydroculture plants.

2. Less guesswork when watering: Watering interior plants (including houseplants) can be a tricky task to master. Soil-based plants have to be watered with a lot of precision and know-how. One of the most common and easiest mistakes made with soil is over-watering. When plants are watered too frequently or too much at one watering, soils become water-logged, depriving the roots of oxygen. This causes the roots to fail, thus causing stress and premature death to the plant.

With hydroculture, it's still possible to over-water or under-water plants but the margin for error is greater. The abundant air present in the LECA enables a stronger root system that is also more forgiving if over-watered.

3. Longer watering cycle: The frequency at which interior plants need to be watered

varies based on many different factors; however, the average watering cycle for a 6" plant is probably every 2 weeks in soil. With hydroculture, the length of time between watering is typically tripled.

A 6" hydroculture plant can typically go six weeks or more without having to be watered again. This can be especially helpful for people that cannot access their plants every two weeks (i.e. away from home, vacation, etc.).

4. Longer-lasting plants: Hydroculture is the ideal growing medium for interior plants because the roots are healthier and more robust. Healthy root systems support longer-lasting plants.

Disadvantages

As with any method of plant cultivation there are disadvantages as well as advantages. One of the biggest disadvantages of using the hydroculture growing method concerns plant stability. In soil, plants can remain stable easily because the medium is dense. In media used for hydroculture, such as expanded clay pellets, plants often have to establish a firm root system before becoming fully stabilised. This means that before a substantial root system has formed any tipping over or forceful movement of plant pots can cause spilling of the growing medium and dislodge the plant, causing unnecessary stress to the plant.

Aside from that the only other real disadvantage with hydroculture is a build-up of salts from fertilisers and chemicals from pest-control products in substrates such as expanded clay pellets. However, maintaining a clean growing environment and using products designed to flush excess fertiliser and chemicals can counteract this issue.

Aeroponics

Aeroponics is an indoor gardening practice in which plants are grown and nourished by suspending their root structures in air and regularly spraying them with a nutrient and water solution.

Soil is not used for aeroponics, because the plants can thrive when their roots are constantly or periodically exposed to a nutrient-rich mist. Aeroponics offers an efficient means to grow plants, including fruits and vegetables, without potting and repotting them to replenish their access to nutrient-rich soil.

The National Aeronautics and Space Administration (NASA) tested the effectiveness of aeroponics on the Mir space station and the results showed that Asian bean seedlings could grow effectively in a nutrient solution in zero gravity. Plants are suspended in the air in enclosed frames that leave the leafy tips and the roots able to grow up and down respectively. Many aeroponic systems look very similar to traditional potted plant systems, with the key difference being that the containers for the plants are sealed around the plants' bases and have a closed environment for the root systems.

Instead of relying on a mixture of soil and water to feed the plants, aeroponic horticulturists spray the root systems with a nutrient mix. Because the roots are enclosed, the nutrient-water mix is used more efficiently by the plants and less water is needed for them to grow and thrive. With aeroponics, indoor horticulturists may use vertical and horizontal space to grow more plants using less floor space and they conserve water by using sealed aeroponic systems.

Depending on the aeroponic system, nutrients may be sprayed manually at intervals throughout the day and night, but most aeroponic systems have one or more pumps that automatically keep plants nourished without constant supervision. As long as the system is sealed and nutrient mist is consistently pumped to the roots, plants should thrive in an aeroponic environment.

Aeroponics uses a small internal microjet spray that sprays the roots with fine, high pressure mist containing nutrient rich solutions. Because the roots are exposed to more oxygen, the plant tends to grow faster. It is also easier to administer all sorts of nutrients to the plant, via the root system.

In a typical aeroponic system, plants are usually suspended on top of a reservoir, within a tightly sealed container. A pump and sprinkler system creates vapors out of a nutrient rich solution, and sprays the result in the reservoir, engulfing the dangling plant roots. Plants are inserted into the platform top holes and supported with collars. Aeroponics is often confused with hydroponics, since the two methods are similar and interchangeable, but In aeroponics the roots have no contact with any media, whereas in hydroponics, they do.

Roots of plants grown via aeroponics.

Some people think that aeroponically grown plants would be more frail than plants grown in the soil, but this is entirely false, as they are in fact even more well fed than most of their soil counterparts. Aeroponics can also be combined perfectly with hydroponics, to produce strong, healthy plants, as in hydro-aeroponics. The secret of aeroponics lies in the increased oxygen available to the roots due to the lack of root zone media.

Commercial aeroponics only took off during the 1980s, but has been growing ever since, because the need is clear – People are always looking for better and more convenient ways to grow plants with a minimum of fuss.

Some of the key benefits of aeroponics:

- Fast plant growth – The chief feature of aeroponics. Plants grow fast because their roots have access to a lot of oxygen 24/7.

- Easy system maintenance – In aeroponics, all you need to maintain is the root chamber (the container housing the roots) which needs regular disinfecting, and periodically, the reservoir and irrigation channels. The constant semi-moist environment of the root chamber which invites bacterial growth is the only main drawback of all aeroponic system maintenance.

- Less need for nutrients and water – Aeroponic plants need less nutrients and water on average, because the nutrient absorption rate is higher, and plants usually respond to aeroponic systems by growing even more roots.

- Mobility – Plants, even whole nurseries, can be moved around without too much effort, as all that is required is moving the plants from one collar to another.

- Requires little space – You don't need much space to start an aeroponics garden. Depending on the system, plants can be stacked up one on top of each other. Aeroponics is basically a modular system, which is perfect for maxing out limited space.

- Great educational value – You can learn a great deal about plants from aeroponics. Kids especially will love having a small aeroponic system to grow a pet plant, without having to get their hands dirty.

Key Disadvantages of Aeroponics

- Dependence on the system – A typical aeroponics system is made up of high pressure pumps, sprinklers and timers. If any of these break down, your plants can be damaged or killed easily.

- Technical knowledge required – You need a certain level of competency in running an aeroponic system. Knowledge of nutrients amounts required by your plant is essential, because you don't have any soil to absorb excess/wrong nutrients supplied.

- Regular cleaning of the root chamber – The root chamber must not be contaminated, or else diseases may strike the roots. So you need to disinfect the root chamber every so often. Hydrogen peroxide is often used as disinfectant.

- High cost – Most aeroponic systems are not exactly cheap. Aeroponic systems may cost many hundreds of dollars each.

Aquaponics

Aquaponics refers to any system that combines conventional aquaculture (raising aquatic animals such as snails, fish, crayfish or prawns in tanks) with hydroponics (cultivating plants in water) in a symbiotic environment. In normal aquaculture, excretions from the animals being raised can accumulate in the water, increasing toxicity. In an aquaponic system, water from an aquaculture system is fed to a hydroponic system where the by-products are broken down by nitrifying bacteria initially into nitrites and subsequently into nitrates that are utilized by the plants as nutrients. Then, the water is recirculated back to the aquaculture system.

As existing hydroponic and aquaculture farming techniques form the basis for all aquaponic systems, the size, complexity, and types of foods grown in an aquaponic system can vary as much as any system found in either distinct farming discipline.

Parts of an Aquaponic System

A commercial aquaponics system. An electric pump moves nutrient-rich water from the fish tank through a solids filter to remove particles the plants above cannot absorb. The water then provides nutrients for the plants and is cleansed before returning to the fish tank below.

Aquaponics consists of two main parts, with the aquaculture part for raising aquatic animals and the hydroponics part for growing plants. Aquatic effluents, resulting from uneaten feed or raising animals like fish, accumulate in water due to the closed-system recirculation of most aquaculture systems. The effluent-rich water becomes toxic to the aquatic animal in high concentrations but this contains nutrients essential for plant growth. Although consisting primarily of these two parts, aquaponics systems are usually grouped into several components or subsystems responsible for the effective removal of solid wastes, for adding bases to neutralize acids, or for maintaining water oxygenation. Typical components include:

- Rearing tank: The tanks for raising and feeding the fish;

- Settling basin: A unit for catching uneaten food and detached biofilms, and for settling out fine particulates;

- Biofilter: A place where the nitrification bacteria can grow and convert ammonia into nitrates, which are usable by the plants;

- Hydroponics subsystem: The portion of the system where plants are grown by absorbing excess nutrients from the water;

- Sump: The lowest point in the system where the water flows to and from which it is pumped back to the rearing tanks.

Depending on the sophistication and cost of the aquaponics system, the units for solids removal, biofiltration, and the hydroponics subsystem may be combined into one unit or subsystem, which prevents the water from flowing directly from the aquaculture part of the system to the hydroponics part. By utilizing gravel or sand as plant supporting medium, solids are captured and the medium has enough surface area for fixed-film nitrification. The ability to combine biofiltration and hydroponics allows for aquaponic system to in many cases eliminate the need for an expensive, separate biofilter.

Live Components

An aquaponic system depends on different live components to work successfully. The three main live components are plants, fish (or other aquatic creatures) and bacteria. Some systems also include additional live components like worms.

Plants

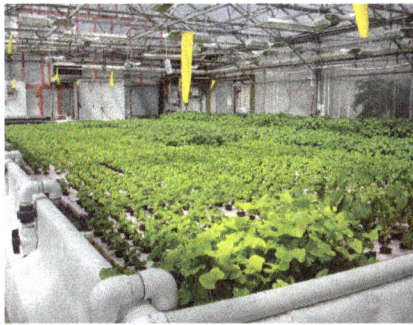

A Deep Water Culture hydroponics system where plant grow directly into the effluent rich water without a soil medium. Plants can be spaced closer together because the roots do not need to expand outwards to support the weight of the plant.

Plant placed into a nutrient rich water channel in a Nutrient film technique (NFT) system.

Many plants are suitable for aquaponic systems, though which ones work for a specific system depends on the maturity and stocking density of the fish. These factors influence the concentration of nutrients from the fish effluent and how much of those nutrients are made available to the plant roots via bacteria. Green leaf vegetables with low to medium nutrient requirements are well adapted to aquaponic systems, including chinese cabbage, lettuce, basil, spinach, chives, herbs, and watercress.

Spinach seedlings, 5 days old, by aquaponics.

Other plants, such as tomatoes, cucumbers, and peppers, have higher nutrient requirements and will do well only in mature aquaponic systems with high stocking densities of fish.

Plants that are common in salads have some of the greatest success in aquaponics, including cucumbers, shallots, tomatoes, lettuce, chiles, capsicum, red salad onions and snow peas.

Some profitable plants for aquaponic systems include chinese cabbage, lettuce, basil, roses, tomatoes, okra, cantaloupe and bell peppers.

Other species of vegetables that grow well in an aquaponic system include watercress, basil, coriander, parsley, lemongrass, sage, beans, peas, kohlrabi, taro, radishes, strawberries, melons, onions, turnips, parsnips, sweet potato, cauliflower, cabbage, broccoli, and eggplant as well as the choys that are used for stir fries.

Fish or Other Aquatic Creatures

Filtered water from the hydroponics system drains into a catfish tank for re-circulation.

Fresh water fish are the most common aquatic animal raised using aquaponics due to their ability to tolerate crowding, although freshwater crayfish and prawns are also sometimes used. There is a branch of aquaponics using saltwater fish, called saltwater aquaponics. There are many species of warmwater and coldwater fish that adapt well to aquaculture systems.

In practice, tilapia are the most popular fish for home and commercial projects that are intended to raise edible fish because it is a warmwater fish species that can tolerate crowding and changing water conditions. Barramundi, silver perch, eel-tailed catfish or tandanus catfish, jade perch and Murray cod are also used. For temperate climates when there isn't ability or desire to maintain water temperature, bluegill and catfish are suitable fish species for home systems.

Koi and goldfish may also be used, if the fish in the system need not be edible. Other suitable fish include channel catfish, rainbow trout, perch, common carp, Arctic char, largemouth bass and striped bass.

Bacteria

Nitrification, the aerobic conversion of ammonia into nitrates, is one of the most important functions in an aquaponic system as it reduces the toxicity of the water for fish, and allows the resulting nitrate compounds to be removed by the plants for nourishment. Ammonia is steadily released into the water through the excreta and gills of fish as a product of their metabolism, but must be filtered out of the water since higher concentrations of ammonia (commonly between 0.5 and 1 ppm) can impair growth, cause widespread damage to tissues, decrease resistance to disease and even kill the fish. Although plants can absorb ammonia from the water to some degree, nitrates are assimilated more easily, thereby efficiently reducing the toxicity of the water for fish. Ammonia can be converted into safer nitrogenous compounds through combined healthy populations of 2 types of bacteria: Nitrosomonas which convert ammonia into nitrites, and Nitrobacter which then convert nitrites into nitrates. While nitrate is still harmful to fish due to its ability to create metehemoglobine, which cannot bind oxygen, by attaching to hemoglobin, nitrates are able to be tolerated at high levels by fish. High surface area provides more space for the growth of nitrifying bacteria. Grow bed material choices require careful analysis of the surface area, price and maintainability considerations.

Hydroponic Subsystem

Plants are grown as in hydroponics systems, with their roots immersed in the nutrient-rich effluent water. This enables them to filter out the ammonia that is toxic to the aquatic animals, or its metabolites. After the water has passed through the hydroponic subsystem, it is cleaned and oxygenated, and can return to the aquaculture vessels. This cycle is continuous. Common aquaponic applications of hydroponic systems include:

- Deep-water raft aquaponics: styrofoam rafts floating in a relatively deep aquaculture basin in troughs. Raft tanks can be constructed to be quite large, and enable seedlings to be transplanted at one end of the tank while fully grown plants are harvested at the other, thus ensuring optimal floor space usage.

- Recirculating aquaponics: solid media such as gravel or clay beads, held in a container that is flooded with water from the aquaculture. This type of aquaponics is also known as closed-loop aquaponics.

- Reciprocating aquaponics: solid media in a container that is alternately flooded and drained utilizing different types of siphon drains. This type of aquaponics is also known as flood-and-drain aquaponics or ebb-and-flow aquaponics.

- Nutrient film technique channels: plants are grown in lengthy narrow channels, with a film of nutrient-filled water constantly flowing past the plant roots. Due to the small amount of water and narrow channels, helpful bacteria cannot live there and therefore a bio filter is required for this method.

- Other systems use towers that are trickle-fed from the top, horizontal PVC pipes with holes for the pots, plastic barrels cut in half with gravel or rafts in them. Each approach has its own benefits.

Since plants at different growth stages require different amounts of minerals and nutrients, plant harvesting is staggered with seedlings growing at the same time as mature plants. This ensures stable nutrient content in the water because of continuous symbiotic cleansing of toxins from the water.

Biofilter

In an aquaponics system, the bacteria responsible for the conversion of ammonia to usable nitrates for plants form a biofilm on all solid surfaces throughout the system that are in constant contact with the water. The submerged roots of the vegetables combined have a large surface area where many bacteria can accumulate. Together with the concentrations of ammonia and nitrites in the water, the surface area determines the speed with which nitrification takes place. Care for these bacterial colonies is important as to regulate the full assimilation of ammonia and nitrite. This is why most aquaponics systems include a biofiltering unit, which helps facilitate growth of these microorganisms. Typically, after a system has stabilized ammonia levels range from 0.25 to.50 ppm; nitrite levels range from 0.0 to 0.25 ppm, and nitrate levels range from 5 to 150 ppm. During system startup, spikes may occur in the levels of ammonia (up to 6.0 ppm) and nitrite (up to 15 ppm), with nitrate levels peaking later in the startup phase. In the nitrification process ammonia is oxidized into nitrite, which releases hydrogen ions into the water. Overtime your pH will slowly drop, so you can use non-sodium bases such as potassium hydroxide or calcium hydroxide to neutralize the water's pH if insufficient quantities are naturally present in the water to provide a buffer against acidification. In addition, selected minerals or nutrients such as iron can be added in addition to the fish waste that serves as the main source of nutrients to plants.

A good way to deal with solids buildup in aquaponics is the use of worms, which liquefy the solid organic matter so that it can be utilized by the plants and other animals in the system. For a worm-only growing method, please see Vermiponics.

Operation

The five main inputs to the system are water, oxygen, light, feed given to the aquatic animals, and electricity to pump, filter, and oxygenate the water. Spawn or fry may be added to replace grown fish that are taken out from the system to retain a stable system. In terms of outputs, an aquaponics system may continually yield plants such as vegetables grown in hydroponics, and edible aquatic species raised in an aquaculture. Typical build ratios are.5 to 1 square foot of grow space for every 1 U.S. gal (3.8 L) of aquaculture water in the system. 1 U.S. gal (3.8 L) of water can support between.5 lb (0.23 kg) and 1 lb (0.45 kg) of fish stock depending on aeration and filtration.

Ten primary guiding principles for creating successful aquaponics systems were issued by Dr. James Rakocy, the director of the aquaponics research team at the University of the Virgin Islands, based on extensive research done as part of the *Agricultural Experiment Station* aquaculture program.

- Use a feeding rate ratio for design calculations,

- Keep feed input relatively constant,

- Supplement with calcium, potassium and iron,

- Ensure good aeration,

- Remove solids,

- Be careful with aggregates,

- Oversize pipes,

- Use biological pest control,

- Ensure adequate biofiltration,

- Control pH.

Feed Source

As in most aquaculture based systems, stock feed often consists of fish meal derived from lower-value species. Ongoing depletion of wild fish stocks makes this practice unsustainable. Organic fish feeds may prove to be a viable alternative that relieves this concern. Other alternatives include growing duckweed with an aquaponics system that feeds the same fish grown on the system, excess worms grown from vermiculture composting, using prepared kitchen scraps, as well as growing black soldier fly larvae to feed to the fish using composting grub growers.

Plant Nutrients

Like hydroponics, a few minerals and micro nutrients can be added to improve plants growth. Iron is the most deficient nutrient in aquaponics, it can be added through mixing Iron Chelate powder with water. Potassium can be added as potassium sulfate through foliar spray. Less vital nutrients include epsom salt, calcium chloride and boron. Biological filtration of aquaculture wastes yield high nitrate concentrations, which is great for leafy greens. For flowering plants with high nutrient demands it is recommended to introduce supplemental nutrients such as magnesium, calcium, potassium, and phosphorus. Common sources are sulfate of potash, potassium bicarbonate, monoammonium phosphate, etc. Nutrient deficiency in wastewater from fish component (RAS) can be completely masked using raw or mineralized sludge, usually containing 3–17 times higher nutrient concentrations. RAS effluents (wastewater and sludge combined) contain adequate N, P, Mg, Ca, S, Fe, Zn, Cu, Ni to meet most aquaponic crop needs. Potassium is generally deficient requiring full-fledged fertilization. Micronutrients B, Mo are partly sufficient and can be easily ameliorated by increasing sludge release. The presumption surrounding 'definite' phyto-toxic sodium levels in RAS effluents should be reconsidered – practical solutions available too. No threat of heavy metal accumulation exists within the aquaponics loop.

Water usage

Aquaponic systems do not typically discharge or exchange water under normal operation, but instead recirculate and reuse water very effectively. The system relies on the relationship between the animals and the plants to maintain a stable aquatic environment that experience a minimum of fluctuation in ambient nutrient and oxygen levels. Plants are able to recover dissolved nutrients from the circulating water, meaning that less water is discharged and the water exchange rate can be minimized. Water is added only to replace water loss from absorption and transpiration by plants, evaporation into the air from surface water, overflow from the system from rainfall, and removal of biomass such as settled solid wastes from the system. As a result, aquaponics uses approximately 2% of the water that a conventionally irrigated farm requires for the same vegetable production. This allows for aquaponic production of both crops and fish in areas where water or fertile land is scarce. Aquaponic systems can also be used to replicate controlled wetland conditions. Constructed wetlands can be useful for biofiltration and treatment of typical household sewage. The nutrient-filled overflow water can be accumulated in catchment tanks, and reused to accelerate growth of crops planted in soil, or it may be pumped back into the aquaponic system to top up the water level.

Energy Usage

Aquaponic installations rely in varying degrees on man-made energy, technological solutions, and environmental control to achieve recirculation and water/ambient temperatures. However, if a system is designed with energy conservation in mind, using alternative energy and a reduced number of pumps by letting the water flow downwards as much as possible, it can be highly energy efficient. While careful design can minimize the risk, aquaponics systems can have multiple 'single points of failure' where problems such as an electrical failure or a pipe blockage can lead to a complete loss of fish stock.

Fish Stocking

In order for aquaponic systems to be financially successful and make a profit whilst also covering its operating expenses, the hydroponic plant components and fish rearing components need to almost constantly be at maximum production capacity. To keep the bio-mass of fish in the system at its maximum (without limiting fish growth), there are 3 main stocking method that can help maintain this maximum.

- Sequential rearing: Multiple age groups of fish share a rearing tank, and when an age group reaches market size they are selectively harvested and replaced with the same amount of fingerlings. Downsides to this method include stressing out the entire pool of fish during each harvest, missing fish resulting in a waste of food/space, and the difficulty of keeping accurate records with frequent harvests.

- Stock splitting: Large quantities of fingerlings are stocked at once and then split into two groups once the tank hits maximum capacity, which is easier to record and eliminates fish being "forgotten". A stress-free way of doing this operation is via "swimways" that connect various rearing tanks and a series of hatches/ moving screens/pumps that move the fish around.

- Multiple rearing units: Entire groups of fish are moved to larger rearing tanks once their current tank hits maximum capacity. Such systems usually have 2–4 tanks that share a filtration system, and when the largest tank is harvested, the other fish groups are each moved up into a bigger tank whilst the smallest tank is restocked with fingerlings. It is also common for there to be several rearing tanks yet no ways to move fish between them, which eliminates the labor of moving fish and allows each tank to be undisturbed during harvesting, even if the space usage is inefficient when the fish are fingerlings.

Ideally the bio-mass of fish in the rearing tanks doesn't exceed 0.5 lbs/gallon, in order to reduce stress from crowding, efficiently feed the fish, and promote healthy growth.

Disease and Pest Management

Although pesticides can normally be used to take care of insect on crops, in an aquaponic system the use of pesticides would threaten the fish ecosystem. On the other hand, if the fish acquire parasites or diseases, therapeutants cannot be used as the plants would absorb them. In order to maintain the symbiotic relationship between the plants and the fish, non-chemical methods such as traps, physical barriers and biological control (such as parasitic wasps/ladybugs to control white flies/aphids) should be used to control pests. The most effective organic pesticide is Neem oil, but only in small quantity to minimize spill over fish's water.

Automation, Monitoring and control

Many have tried to create automatic control and monitoring systems and some of these demonstrated a level of success. For instance, researchers were able to introduce automation in a small scale aquaponic system to achieve a cost-effective and sustainable farming system. Commercial development of automation technologies has also emerged. For instance, a company has developed a system capable of automating the repetitive tasks of farming and features a machine learning algorithm that can automatically detect and eliminate diseased or underdeveloped plants. A 3.75-acre aquaponics facility that claims to be the first indoor salmon farm in the United States also includes an automated technology. The aquaponic machine has made notable strides in the documenting and gathering of information regarding aquaponics.

Economic Viability

Aquaponics offers a diverse and stable polyculture system that allows farmers to grow vegetables and raise fish at the same time. By having two sources of profit, farmers can continue to earn money even if the market for either fish or plants goes through a low cycle. The flexibility of an aquaponic system allows it to grow a large variety of crops including ordinary vegetables, herbs, flowers and aquatic plants to cater to a broad spectrum of consumers. Herbs, lettuce and speciality greens such as basil or spinach are especially well suited for aquaponic systems due to their low nutritional needs. For the growing number of environmentally conscious consumers, products from aquaponic systems are organic and pesticide free, whilst also leaving a small environmental footprint. Aquaponic systems additionally are economically efficient due to low water usage, effective nutrient cycling and needing little land to operate. Because soil isn't needed and only a little bit of water is required, aquaponic systems can be set up in areas that have traditionally poor soil quality or contaminated water. More importantly, aquaponic systems are usually free of weeds, pests and diseases that would affect soil, which allows them to consistently and quickly produce high quality crops to sell.

Current Examples

The Caribbean island of Barbados created an initiative to start aquaponics systems at home, called the aquaponic machine, with revenue generated by selling produce to tourists in an effort to reduce growing dependence on imported food.

Dakota College at Bottineau in Bottineau, North Dakota has an aquaponics program that gives students the ability to obtain a certificate or an AAS degree in aquaponics.

Vegetable production part of the low-cost Backyard Aquaponics System
developed at Bangladesh Agricultural University

In Bangladesh, the world's most densely populated country, most farmers use agro-chemicals to enhance food production and storage life, though the country lacks over-sight on safe levels of chemicals in foods for human consumption. To combat this issue, a team led by M. A. Salam at the Department of Aquaculture of Bangladesh Agricultural University has created plans for a low-cost aquaponics system to pro-vide organic produce and fish for people living in adverse climatic conditions such as the salinity-prone southern area and the flood-prone haor area in the eastern region. Salam's work innovates a form of subsistence farming for micro-production goals at the community and personal levels whereas design work by Chowdhury and Graff was aimed exclusively at the commercial level, the latter of the two approaches take advan-tage of economies of scale.

With more than a third of Palestinian agricultural lands in the Gaza Strip turned into a buffer zone by Israel, an aquaponic gardening system is developed appropriate for use on rooftops in Gaza City.

The Smith Road facility in Denver started pilot program of aquaponics to feed 800 to 1000 inmates at Denver Jail and neighboring downtown facility which consist of 1,500 inmates and 700 officers. In Malaysia Alor Gajah, Melaka, Organization 'Persatuan Akuakutur Malaysia' takes innovative approach in aquaponics by growing Lobster in aquaponics.

VertiFarms in New Orleans targets corporate rooftops for vertical farming, accruing up to 90 corporate clients for rooftop vertical farming in 2013. Windy Drumlins Farm in Wisconsin redesigns aquaponic-solar greenhouse for extreme weather conditions which can endure extremely cold climate. Volunteer operation in Nicaragua "Amigos for Christ" manages its plantation for feeding 900+ poverty-stricken school children by using nutrients from aquaponics method.

Aquaponics in India aims to provide aspiring farmers with aquaponics solutions for commercial and backyard operation.

Verticulture in Bedstuy utilizes old Pfizer manufacturing plant for producing basil in commercial scale through aquaponics, yielding 30-40 pounds of basil a week.

Aquaponics startup Edenworks in New York expands to full-scale commercial facility, which will generate 130,000 pounds of greens and 50,000 pounds of fish a year.

There has been a shift towards community integration of aquaponics, such as the non-profit foundation Growing Power that offers Milwaukee youth job opportunities and training while growing food for their community. The model has spawned several satellite projects in other cities, such as New Orleans where the Vietnamese fisherman community has suffered from the Deepwater Horizon oil spill, and in the South Bronx in New York City.

Whispering Roots is a non-profit organization in Omaha, Nebraska that provides fresh, locally grown, healthy food for socially and economically disadvantaged communities by using aquaponics, hydroponics and urban farming.

In addition, aquaponic gardeners from all around the world are gathering in online community sites and forums to share their experiences and promote the development of this form of gardening as well as creating extensive resources on how to build home systems.

Recently, aquaponics has been moving towards indoor production systems. In cities like Chicago, entrepreneurs are utilizing vertical designs to grow food year round. These systems can be used to grow food year round with minimal to no waste.

There are various modular systems made for the public that utilize aquaponic systems to produce organic vegetables and herbs, and provide indoor decor at the same time. These systems can serve as a source of herbs and vegetables indoors. Universities are promoting research on these modular systems as they get more popular among city dwellers.

Hydroponics

Hydroponics is a method of growing plants without using soil (i.e., soil less). This technique instead uses a mineral nutrient solution in a water solvent, allowing the nutrient uptake process to be more efficient than when using soil. There are several types or variations of hydroponics.

Hydroponics is suitable for commercial food producers and hobbyist gardeners alike. Hydroponics possesses several advantages over a soil medium. Unlike plants grown in soil, plants grown in a hydroponics system do not need to develop extensive root structures to search for nutrients. It is easier to test and adjust pH levels. In the hydroponics method, plants are raised in an inert and perfectly pH balanced growing medium where the plants only need to expend minimal energy to acquire nutrients from the roots. The energy saved by the roots is better spent on fruit and flower production.

There are several types of hydroponic growing techniques, including:

- Nutrient film technique (NFT),

- Wicks system,

- Ebb and flow (flood and drain),

- Water culture,

- Drip system,

- Aeroponic system.

In soils, nutrients and water are randomly placed, and often plants need to expend a lot of energy to find the water and nutrients by growing roots to find them. By expending this energy, the plants growth is not as fast as it could be. In a hydroponic garden, the nutrients and water are delivered straight to the plants roots, allowing the plants to grow faster, and allowing harvesting to be done sooner, simply because the plants are putting more of their energy into growing above the ground, instead of under it.

Once a plant is established it gives higher than average yields, whether being grown in a greenhouse, a backyard or a balcony. Also, hydroponics allows you to grow more plants per square metre. This is because the plants do not need to compete with weeds and each other for the food and water that is in soil, this food and water is delivered straight to them.

It is also very important to note that, despite many myths, plants grown in hydroponics are no different to plants grow in soil, they will have the same physiology. Plants grown in a hydroponic system take the same nutrients as those grown in soil, though the content can be more accurately controlled. The basic difference between the two methods is the way in which nutrients and water are delivered to the plants.

In hydroponics, the nutrient salts are already refined and the plants do not need to wait for the nutrients to break down to their basic form. However, with soil based agriculture, plants are fed nutrients via manures and composts which must break down into their basic form (nutrient salts) before the plants can use them.

Advantages of Hydroponics

No Soils Needed

In a sense, you can grow crops in places where the land is limited, doesn't exist, or is heavily contaminated. In the 1940s, Hydroponics was successfully used to supply fresh vegetables for troops in Wake Island, a refueling stop for Pan American airlines. This is a distant arable area in the Pacific Ocean. Also, Hydroponics has been considered as the farming of the future to grow foods for astronauts in the space (where there is no soil) by NASA.

Make Better use of Space and Location

Because all that plants need are provided and maintained in a system, you can grow in your small apartment, or the spare bedrooms as long as you have some spaces.

Plants' roots usually expand and spread out in search of foods, and oxygen in the soil. This is not the case in Hydroponics, where the roots are sunk in a tank full of oxygenated nutrient solution and directly contact with vital minerals. This means you can grow your plants much closer, and consequently huge space savings.

Climate Control

Like in greenhouses, hydroponic growers can have total control over the climate - temperature, humidity, light intensification, the composition of the air. In this sense, you can grow foods all year round regardless of the season. Farmers can produce foods at the appropriate time to maximize their business profits.

Hydroponics is Water-saving

Plants grown hydroponically can use only 10% of water compared to field-grown ones. In this method, water is recirculated. Plants will take up the necessary water, while run-off ones will be captured and return to the system. Water loss only occurs in two forms - evaporation and leaks from the system (but an efficient hydroponic setup will minimize or don't have any leaks).

It is estimated that agriculture uses up to 80% water of the ground and surface water in the US.

While water will become a critical issue in the future when food production is predicted to increase by 70% according to the FAQ, Hydroponics is considered a viable solution to large-scale food production.

Effective use of Nutrients

In Hydroponics, you have a 100% control of the nutrients (foods) that plants need. Before planting, growers can check what plants require and the specific amounts of nutrients needed at particular stages and mix them with water accordingly. Nutrients are conserved in the tank, so there are no losses or changes of nutrients like they are in the soil.

pH Control of the Solution

All of the minerals are contained in the water. That means you can measure and adjust the pH levels of your water mixture much more easily compared to the soils. That ensures the optimal nutrients uptake for plants.

Better Growth Rate

Hydroponically plants grow faster than in soil.

Plants are placed in ideal conditions, while nutrients are provided at the sufficient amounts, and come into direct contacts with the root systems. Thereby, plants no longer waste valuable energy searching for diluted nutrients in the soil. Instead, they shift all of their focus on growing and producing fruits.

No Weeds

If you have grown in the soil, you will understand how irritating weeds cause to your garden. It's one of the most time-consuming tasks for gardeners - till, plow, hoe, and so on. Weeds are mostly associated with the soil. So eliminate soils, and all bothers of weeds are gone.

Fewer Pests and Diseases

And like weeds, getting rids of soils helps make your plants less vulnerable to soil-borne pests like birds, gophers, groundhogs; and diseases like Fusarium, Pythium, and Rhizoctonia species.Also when growing indoors in a closed system, the gardeners can easily take controls of most surrounding variables.

Less use of Insecticide and Herbicides

Since you are using no soils and while the weeds, pests, and plant diseases are heavily reduced, there are fewer chemicals used. This helps you grow cleaner and healthier foods. The cut of insecticide and herbicides is a strong point of Hydroponics when the criteria for modern life and food safety are more and more placed on top.

Labor and Time Savers

Besides spending fewer works on tilling, watering, cultivating, and fumigating weeds and pests, you enjoy much time saved because plants' growth is proven to be higher in Hydroponics. When agriculture is planned to be more technology-based, Hydroponics has a room in it.

The disadvantages of hydroponics are:

Experiences and Technical Knowledge

You are running a system of many types of equipment, which requires necessary specific expertise for the devices used, what plants you can grow and how they can survive and thrive in a soilless environment. Mistakes in setting up the systems and plants' growth ability in this soilless environment and you end up ruining your whole progress.

Water and Electricity Risks

In a Hydroponic system, mostly you use water and electricity. Beware of electricity in a combination of water in close proximity. Always put safety first when working with the water systems and electric equipment, especially in commercial greenhouses.

System Failure Threats

You are using electricity to manage the whole system. So suppose you do not take preliminary actions for a power outage, the system will stop working immediately, and plants may dry out quickly and will die in several hours. Hence, a backup power source and plan should always be planned, especially for great scale systems.

Initial Expenses

You are sure to spend under one hundred to a few hundreds of dollars (depending on your garden scale) to purchase equipment for your first installation. Whatever systems you build, you will need containers, lights, a pump, a timer, growing media, nutrients). Once the system has been in place, the cost will be reduced to only nutrients and electricity (to keep the water system running, and lighting).

Long Return Per Investment

If you follow news on agriculture start-up, you may have known that there have been some new indoor hydroponic business started recently. That's a good thing for the agriculture sector and the development of Hydroponics as well. However, commercial growers still face some big challenges when starting with Hydroponics on a large scale. This is largely because of the high initial expenses and the long, uncertain ROI (return on investment). It's not easy to detail a clear profitable plan to urge for investment while there are also many other attractive high-tech fields out there that seem fairly promising for funding.

Diseases and Pests may Spread Quickly

You are growing plants in a closed system using water. In the case of plant infections or pests, they can escalate fast to plants on the same nutrient reservoir. In most cases, diseases and pests are not so much of problem in a small system of home growers.

Regenerative Agriculture

Regenerative Agriculture describes farming and grazing practices that, among other benefits, reverse climate change by rebuilding soil organic matter and restoring degraded soil biodiversity – resulting in both carbon drawdown and improving the water cycle.

Specifically, Regenerative Agriculture is a holistic land management practice that leverages the power of photosynthesis in plants to close the carbon cycle, and build soil health, crop resilience and nutrient density. Regenerative agriculture improves soil health, primarily through the practices that increase soil organic matter. This not only aids in increasing soil biota diversity and health, but increases biodiversity both above and below the soil surface, while increasing both water holding capacity and sequestering carbon at greater depths, thus drawing down climate-damaging levels of atmospheric CO_2, and improving soil structure to reverse civilization-threatening human-caused soil loss. Research continues to reveal the damaging effects to soil from tillage, applications of agricultural chemicals and salt based fertilizers, and carbon mining. Regenerative Agriculture reverses this paradigm to build for the future.

Regenerative Agricultural Practices are:

Practices that (i) contribute to generating/building soils and soil fertility and health; (ii) increase water percolation, water retention, and clean and safe water runoff; (iii) increase biodiversity and ecosystem health and resiliency; and (iv) invert the carbon emissions of our current agriculture to one of remarkably significant carbon sequestration thereby cleansing the atmosphere of legacy levels of CO_2.

Practices Include

1. No-till/minimum tillage. Tillage breaks up (pulverizes) soil aggregation and fungal communities while adding excess O_2 to the soil for increased respiration and CO_2 emission. It can be one of the most degrading agricultural practices, greatly increasing soil erosion and carbon loss. A secondary effect is soil capping and slaking that can plug soil spaces for percolation creating much more water runoff and soil loss. Conversely, no-till/minimum tillage, in conjunction with other regenerative practices, enhances soil aggregation, water infiltration and retention, and carbon sequestration. However, some soils benefit from interim ripping to break apart hardpans, which can increase root zones and yields and have the capacity to increase soil health and carbon sequestration. Certain low level chiseling may have similar positive effects.

2. Soil fertility is increased in regenerative systems biologically through application of cover crops, crop rotations, compost, and animal manures, which restore the plant/soil microbiome to promote liberation, transfer, and cycling of essential soil nutrients. Artificial and synthetic fertilizers have created imbalances in the structure and function of microbial communities in soils, bypassing the natural biological acquisition of nutrients for the plants, creating a dependent agroecosystem and weaker, less resilient plants. Research has observed that application of synthetic and artificial fertilizers contribute to climate change through (i) the energy costs of production and transportation of the fertilizers, (ii) chemical breakdown and migration into water resources and the atmosphere; (iii) the

distortion of soil microbial communities including the diminution of soil methanothrops, and (iv) the accelerated decomposition of soil organic matter.

3. Building biological ecosystem diversity begins with inoculation of soils with composts or compost extracts to restore soil microbial community population, structure and functionality restoring soil system energy (Ccompounds as exudates) through full-time planting of multiple crop intercrop plantings, multispecies cover crops, and borders planted for bee habitat and other beneficial insects. This can include the highly successful push-pull systems. It is critical to change synthetic nutrient dependent monocultures, low-biodiversity and soil degrading practices.

4. Well-managed grazing practices stimulate improved plant growth, increased soil carbon deposits, and overall pasture and grazing land productivity while greatly increasing soil fertility, insect and plant biodiversity, and soil carbon sequestration. These practices not only improve ecological health, but also the health of the animal and human consumer through improved micro-nutrients availability and better dietary omega balances. Feed lots and confined animal feeding systems contribute dramatically to (i) unhealthy monoculture production systems, (ii) low nutrient density forage (iii) increased water pollution, (iv) antibiotic usage and resistance, and (v) CO_2 and methane emissions, all of which together yield broken and ecosystem-degrading food-production systems.

Integrated Pest Management

Integrated pest management (IPM) is an eco-friendlier approach to pest control than traditional pesticide use. Instead of simply spraying plants with pesticides to keep insects and other vermin away, IPM focuses on preventing pests from infesting plants in the first place, and uses pesticides only when it is absolutely necessary.

Integrated pest management may also be known as integrated pest control (IPC).

Integrated pest management is a more environmentally friendly and sustainable pest control solution than traditional pesticide use. It specifically targets pests and deals with them in a way that will not harm beneficial organisms and will minimize pesticide exposure.

The philosophy of integrated pest management is, essentially, that the dangers of pesticide exposure are far greater than the benefits of using pesticides as a preventive measure against pests. In many cases, non-chemical means of pest prevention are available that have the same – if not better – results without the risk of exposing people or animals to harmful chemicals.

As its name suggests, IPM involves the integration of multiple pest control methods, based on site-specific information. This information is gathered through inspection, monitoring, and updated reports, which can inform gardeners, botanists, and horticulturists of the best pest prevention methods.

By identifying the specific pests that are a problem and monitoring your progress with them, you can determine the best measures to take to prevent those pests from harming your plants. This will help you reduce or even eliminate pesticide use without allowing your plants to suffer.

You can create an action plan to prevent pests infestations, including non-chemical measures to discourage pests (e.g., reducing clutter), sealing off the area so that pests cannot access it, and other measures. After taking preventive measures, you can then take pest control measures to reduce the pest population that is still reaching and harming your plants.

Importance of IPM

You might be wondering why you should even consider IPM when chemical pesticides so often succeed at controlling pests. Here are some reasons for having a broader approach to pest management than just the use of chemicals.

- Keep a Balanced Ecosystem: Every ecosystem, made up of living things and their non-living environment, has a balance; the actions of one creature in the ecosystem usually affect other, different organisms. The introduction of chemicals into the ecosystem can change this balance, destroying certain species and allowing other species (sometimes pests themselves) to dominate. Beneficial insects such as the ladybird beetle and lacewing larvae, both of which consume pests, can be killed by pesticides, leaving few natural mechanisms of pest control.

- Pesticides Can be Ineffective: Chemical pesticides are not always effective. Pests can become resistant to pesticides. In fact, some 600 cases of pests developing pesticide resistance have been documented to date, including common lamb's-quarter, house flies, the Colorado potato beetle, the Indian meal moth, Norway rats, and the greenhouse whitefly. Furthermore, pests may survive in some situations where the chemical does not reach pests, is washed off, is applied at an improper rate, or is applied at an improper life stage of the pest.

- IPM Is Not Difficult: Although some of the terms and ideas may be new to you, practicing IPM is not difficult. Believe it or not, you will have done much of the "work" for an IPM approach if you've figured out the problem (the pest), determined the extent of the damage, and decided on the action to take. These steps are the same ones used in IPM.

- Save Money: IPM can save money through avoiding crop loss (due to pests), and avoiding unnecessary pesticide expense.

- Promote a Healthy Environment: We have much to learn about the persistence of chemicals in the environment, and their effect on living creatures. However, more cases of contaminated groundwater appear each year, and disposal of containers and unused pesticides still pose challenges for applicators. Even though long-term documentation on the effects of all pesticides is still unavailable, it is generally agreed that fewer pesticides means less risk to surface water and groundwater, and less hazard to wildlife and humans.

- Maintain a Good Public Image: Recent public outcry about the use of growth regulators and the presence of pesticide residues on produce has heightened pesticide applicator awareness of the level of public concern about chemicals. Consumers are pressuring food stores, which in turn are pressuring producers, for produce that has been grown with as few pesticides as possible. Growing food under integrated pest management can help allay public concerns. Structural pest control professionals can suggest improvements in housekeeping or structural modifications as substitutes for chemical control.

The Basic Steps of IPM

All of the components of an IPM approach can be grouped into four major steps. The first step is taking preventative measures to prevent pest buildup, the second is monitoring, the third step is assessing the pest situation, and the fourth is determining the best action to take.

Preventative Measures

Many IPM practices are used before a pest problem develops to prevent or stall the buildup of pests.

- Cultural Controls: Are those that disrupt the environment of the pest. Plowing, crop rotation, removal of infected plant material, sanitation of greenhouse equipment, and effective manure management are all cultural practices that are employed to deprive pests of a comfortable habitat. The management of urban and industrial pests has improved when sanitation programs have been improved, pest harborages eliminated, garbage pickup frequency increased, or when lights are installed that do not attract insects.

- Structural Modifications: By preventing support timbers from soil contact, damage from several different wood destroying pests can be avoided. Wood absorbs moisture and is more susceptible to attack by carpenter ants and termites when in direct contact with the soil.

- Construction Site Sanitation: Removing tree stumps and lumber scraps from construction sites, which are prime food sources for subterranean termites, can prevent problems in the future.

- Biological Controls - using natural enemies (biological control agents) to keep pests in check can be put into place before pest problems increase. Examples of biological control agents are beneficial mites that feed on mite pests in orchards, the milky spore disease that kills harmful soil grubs, and Encarsia formosa, a wasp that parasitizes the greenhouse whitefly. Many biological control agents are commercially available.

- Physical Barriers such as netting over small fruits and screening in greenhouses can prevent crop loss. Physical barriers are important in termite, house fly, and rodent control.

- Use of Pheromones (natural insect scents) has become widely used in pest management. Sometimes a manufactured "copy" of the pheromone that a female insect emits to attract mates can be used to confuse males and prevent mating. This technique is used in curbing damage from the grape berry moth.

- Pest-Resistant Varieties are those that are less susceptible than other varieties to certain insects and diseases. Use of resistant varieties often means that growers do not need to apply as many pesticides as with susceptible varieties. Potato growers control the golden nematode by planting resistant cultivars. Apple growers can save up to eight fungicide applications a year by growing Liberty and Freedom cultivars, which resist diseases. Farmers growing alfalfa and wheat keep several pests at bay by planting resistant varieties.

Once a pest manager has taken precautions to prevent pest infestations, it is important to watch regularly for the appearance of insects, weeds, diseases, and other pests.

Monitoring (Scouting)

Monitoring pests involves:

- Regular checking of the area;

- Early detection of pests;

- Proper identification of pests;

- Identification of the effects of biological control agents.

  - Regular checking of a warehouse, bakery, restaurant, field, greenhouse, golf course, or other areas and early detection of pests can function together like an early warning system for pests, helping to avoid or prevent a pest problem.

- Proper identification of pests is an extremely important prerequisite to handling problems effectively. For example, the brown banded and German cockroach can be easily confused with each other. Identification is important because certain management practices may control only one species and not the other. Correct identification enables you to manage the real source of the problem and avoid merely treating the symptoms (or controlling non-pests). Some pests cause similar evidence. Unless the pest is identified, the control program may have the wrong pest as its target. Identification enables you to cure the pest problem and avoid injury to non-target organisms, particularly if you:

  - Use a pesticide that is specific to the pest;

  - Control the pest effectively during the most susceptible stage of its life cycle;

  - Consider the use of a non-chemical control.

- Identifying the effects of biological control means knowing which creatures are helpful and determining if pests have been affected by the beneficial organisms. Sometimes pests are kept in check naturally, and at other times the pest populations increase sharply.

Assessment

Assessment is the process of determining the potential for pest populations to reach an economic threshold or an intolerable level. Is a grower likely to suffer financially? Is the pest likely to transmit a disease? How can you tell? There are important differences between the assessment of crop pests and urban pests.

- Forecasting: Can help you determine if weather conditions will be favorable for the development of diseases and insect pests. For example, by "plugging in" values (such as the number of rainy days and the temperatures for those days), growers can predict outbreaks and spray only when conditions are favorable for diseases. Growers who have kept good records of pests in previous years can use these records to help determine if problems such as weeds, insects, and diseases will reoccur. They might be able, for example, to apply the most effective herbicides at the proper time for early control of a problem.

- Thresholds: Or more specifically economic thresholds, are levels that mark the highest point a pest population can reach without risk of economic loss. Populations above these thresholds can reach the economic injury level, where they cause enough damage for the grower to lose money. At the economic injury level, the cost of control is equal to the loss of yield or quality that would result otherwise.

Thresholds for many pests and crops have been scientifically determined. The advantage of thresholds is that if a pest has not reached threshold, there is no risk of economic loss. Therefore, there is no need to spray. Once the pest density (number of pests per unit area) has reached threshold, action is justified. The costs of control will be less than equal to the estimated losses that the pests would cause if left uncontrolled.

Urban pest thresholds are often related to aesthetics rather than economic considerations. Where health concerns or individual sensitivities exist, the tolerable level of the pest may be zero. A zero threshold forces action, even if only one pest has been detected. Zero thresholds exist in hospitals, food production, warehousing, and retail facilities.

Action (Control Measures)

Once a pest has reached the economic threshold, or intolerable level, action should be taken. In some situations, cultural controls can destroy pests. One example is early harvesting to avoid pest problems, which prevents crop loss and can sometimes be more economical than a pesticide application.

Chemical pesticides are used as a control measure when no other strategies will bring the pest population under the threshold. In fact, the success of waiting until a pest reaches threshold usually hinges on the availability of a pesticide that will bring the pest populations down quickly.

In summary, an IPM approach means that pest managers use multiple tactics to prevent pest buildups, monitor pest populations, assess the damage, and make informed management decisions, keeping in mind that pesticides should be used judiciously.

Biological Pest Control

Biological control, biocontrol, or biological pest control is a method of suppressing or controlling the population of undesirable insects, other animals, or plants by the introduction, encouragement, or artificial increase of their natural enemies to economically non–important levels. It is an important component of integrated pest management (IPM) programs.

The biological control of pests and weeds relies on predation, parasitism, herbivory, or other natural mechanisms. Therefore, it is the active manipulation of natural phenomena in serving human purpose, working harmoniously with nature. A successful story of biological control of pests refer to the human beings' capability to depict natural processes for their use and can be the most harmless, non–polluting, and self–perpetuating control method.

In biological control, the reduction of pest populations is achieved by actively using natural enemies.

Natural enemies of the pests, also known as biological control agents, include predatory and parasitoidal insects, predatory vertebrates, nematode parasites, protozoan parasites, and fungal, bacterial, as well as viral pathogens. Biological control agents of plant diseases are most often referred to as antagonists. Biological control agents of weeds include herbivores and plant pathogens. Predators, such as lady beetles and lacewings, are mainly free–living species that consume a large number of prey during their lifetime. Parasitoids are species whose immature stage develops on or within a single insect host, ultimately killing the host. Most have a very narrow host range. Many species of wasps and some flies are parasitoids. Pathogens are disease–causing organisms including bacteria, fungi, and viruses. They kill or debilitate their host and are relatively specific to certain pest or weed groups.

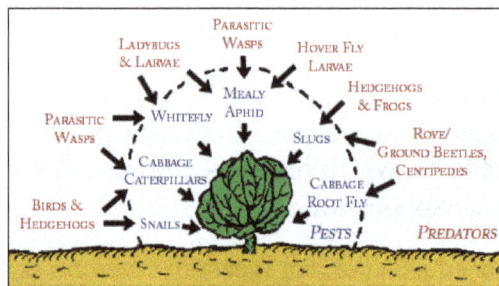

Diagram illustrating the natural enemies of cabbage pests

Strategies of Biological Control Methods

There are three basic types of biological control strategies; conservation biocontrol, classical biological control, and augmentative biological control (biopesticides).

Conservation Biocontrol

The conservation of existing natural enemies is probably the most important and readily available biological control practice available to homeowners and gardeners. Natural enemies occur in all areas, from the backyard garden to the commercial field. They are adapted to the local environment and to the target pest, and their conservation is generally simple and cost–effective. For example, snakes consume a lot or rodent and insect pests that can be damaging to agricultural crops or spread disease. Dragonflies are important consumers of mosquitoes.

Eggs, larvae, and pupae of Helicoverpa moths, the main insect pests of cotton, are all attacked by many beneficial insects and research can be conducted in identifying critical habitats, resources needed to maintain them, and ways of encouraging their activity. Lacewings, lady beetles, hover fly larvae, and parasitized aphid mummies are almost always present in aphid colonies. Fungus–infected adult flies are often common following periods of high humidity. These naturally occurring biological controls are often susceptible to the same pesticides used to target their hosts. Preventing the accidental eradication of natural enemies is termed simple conservation.

Classical Biological Control

Classical biological control is the introduction of exotic natural enemies to a new locale where they did not originate or do not occur naturally. This is usually done by government authorities.

In many instances, the complex of natural enemies associated with an insect pest may be inadequate. This is especially evident when an insect pest is accidentally introduced into a new geographic area without its associated natural enemies. These introduced pests are referred to as exotic pests and comprise about 40 percent of the insect pests in the United States. Examples of introduced vegetable pests include the European corn borer, one of the most destructive insects in North America.

To obtain the needed natural enemies, scientists have utilized classical biological control. This is the practice of importing, and releasing for establishment, natural enemies to control an introduced (exotic) pest, although it is also practiced against native insect pests. The first step in the process is to determine the origin of the introduced pest and then collect appropriate natural enemies associated with the pest or closely related species. The natural enemy is then passed through a rigorous quarantine process, to ensure that no unwanted organisms (such as hyperparasitoids or parasites of the parasite) are introduced, then they are mass produced, and released. Follow–up studies are conducted to determine if the natural enemy becomes successfully established at the site of release, and to assess the long–term benefit of its presence.

There are many examples of successful classical biological control programs. One of the earliest successes was with the cottony cushion scale (Icerya purchasi), a pest that was devastating the California citrus industry in the late 1800s. A predatory insect, the Australian lady beetle or vedalia beetle (Rodolia cardinalis), and a parasitoid fly were introduced from Australia. Within a few years, the cottony cushion scale was completely controlled by these introduced natural enemies. Damage from the alfalfa weevil, a serious introduced pest of forage, was substantially reduced by the introduction of several natural enemies like imported ichnemonid parasitoid Bathyplectes curculionis. About twenty years after their introduction, the alfalfa area treated for alfalfa weevil in the northeastern United States was reduced by 75 percent. A small wasp, Trichogramma ostriniae, introduced from China to help control the European corn borer (Pyrausta nubilalis), is a recent example of a long history of classical biological control efforts for this major pest. Many classical biological control programs for insect pests and weeds are under way across the United States and Canada.

Classical biological control is long lasting and inexpensive. Other than the initial costs of collection, importation, and rearing, little expense is incurred. When a natural enemy is successfully established it rarely requires additional input and it continues to kill the pest with no direct help from humans and at no cost. Unfortunately, classical biological control does not always work. It is usually most effective against exotic pests and less so against native insect pests. The reasons for failure are often not known, but may include the release of too few individuals, poor adaptation of the natural enemy to environmental conditions at the release location, and lack of synchrony between the life cycle of the natural enemy and host pest.

Augmentative Biological Control

This third strategy of biological control method involves the supplemental release of natural enemies. Relatively few natural enemies may be released at a critical time of the season (inoculative release) or literally millions may be released (inundative release). Additionally, the cropping system may be modified to favor or augment the natural enemies. This latter practice is frequently referred to as habitat manipulation.

An example of inoculative release occurs in greenhouse production of several crops. Periodic releases of the parasitoid, Encarsia formosa, are used to control greenhouse whitefly, and the predaceous mite, Phytoseilus persimilis, is used for control of the two–spotted spider mite. The wasp Encarsia formosa lays its eggs in young whitefly "scales," turning them black as the parasite larvae pupates. Ideally it is introduced as soon as possible after the first adult whitefly are seen. It is most effective when dealing with low level infestations, giving protection over a long period of time. The predatory mite, Phytoseilus persimilis, is slightly larger than its prey and has an orange body. It develops from egg to adult twice as fast as the red spider mite and once established quickly overcomes infestation.

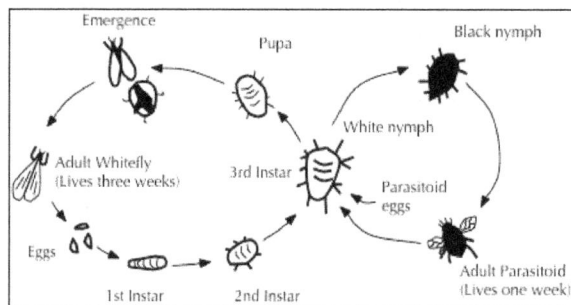

Diagram illustrating the life cycles of Greenhouse whitefly
and its parasitoid wasp *Encarsia formosa*

Lady beetles, lacewings, or parasitoids such as Trichogramma are frequently released in large numbers (inundative release) and are often known as biopesticides. Recommended release rates for Trichogramma in vegetable or field crops range from 5,000 to 200,000 per acre per week depending on level of pest infestation. Similarly, entomoparasitic nematodes are released at rates of millions and even billions per acre for control of certain soil-dwelling insect pests. Entomopathogenic fungus Metarhizium anisopliae var. acridum, which is specific to species of short–horned grasshoppers (Acridoidea and Pyrgomorphoidea) widely distributed in Africa, has been developed as inundative biological control agent.

Habitat or environmental manipulation is another form of augmentation. This tactic involves altering the cropping system to augment or enhance the effectiveness of a natural enemy. Many adult parasitoids and predators benefit from sources of nectar and the protection provided by refuges such as hedgerows, cover crops, and weedy borders. Mixed plantings and the provision of flowering borders can increase the diversity of habitats and provide shelter and alternative food sources. They are easily incorporated into home gardens and even small-scale commercial plantings, but are more difficult to accommodate in large–scale crop production. There may also be some conflict with pest control for the large producer because of the difficulty of targeting the pest species and the use of refuges by the pest insects as well as natural enemies.

Examples of habitat manipulation include growing flowering plants (pollen and nectar sources) near crops to attract and maintain populations of natural enemies. For example, hover fly adults can be attracted to umbelliferous plants in bloom.

Biological control experts in California have demonstrated that planting prune trees in grape vineyards provides an improved overwintering habitat or refuge for a key grape pest parasitoid. The prune trees harbor an alternate host for the parasitoid, which could previously overwinter only at great distances from most vineyards. Caution should be used with this tactic because some plants attractive to natural enemies may also be hosts for certain plant diseases, especially plant viruses that could be vectored by insect pests to the crop. Although the tactic appears to hold much promise, only a few examples have been adequately researched and developed.

Different Types of Biological Control Agents

Predators

Ladybird larva eating wooly apple aphids

Lacewings are available from biocontrol dealers.

Ladybugs, and in particular their larvae which are active between May and July in the northern hemisphere, are voracious predators of aphids such as greenfly and blackfly, and will also consume mites, scale insects, and small caterpillars. The ladybug is a very familiar beetle with various colored markings, while its larvae are initially small and spidery, growing up to 17 millimeters (mm) long. The larvae have a tapering segmented gray/black body with orange/yellow markings nettles in the garden and by leaving hollow stems and some plant debris over–winter so that they can hibernate over winter.

Hoverflies, resembling slightly darker bees or wasps, have characteristic hovering, darting flight patterns. There are over 100 species of hoverfly, whose larvae principally feed upon greenfly, one larva devouring up to 50 a day, or 1000 in its lifetime. They also eat fruit tree spider mites and small caterpillars. Adults feed on nectar and pollen, which they require for egg production. Eggs are minute (1 mm), pale yellow-white, and laid singly near greenfly colonies. Larvae are 8–17 mm long, disguised to resemble bird droppings; they are legless and have no distinct head. Therefore, they are semi–transparent with a range of colors from green, white, brown, and black. Hoverflies can be encouraged by growing attractant flowers such as the poached eggplant (Limnanthes douglasii), marigolds, or phacelia throughout the growing season.

Dragonflies are important predators of mosquitoes, both in the water, where the dragonfly naiads eat mosquito larvae, and in the air, where adult dragonflies capture and eat adult mosquitoes. Community–wide mosquito control programs that spray adult mosquitoes also kill dragonflies, thus removing an important biocontrol agent, and can actually increase mosquito populations in the long term.

Other useful garden predators include lacewings, pirate bugs, rove and ground beetles, aphid midge, centipedes, as well as larger fauna such as frogs, toads, lizards, hedgehogs, slow–worms, and birds. Cats and rat terriers kill field mice, rats, june bugs, and birds. Dogs chase away many types of pest animals. Dachshunds are bred specifically to fit inside tunnels underground to kill badgers.

Parasitoidal Insects

Most insect parasitoids are wasps or flies. For example, the parasitoid Gonatocerus ashmeadi (Hymenoptera: Mymaridae) has been introduced to control the glassy-winged sharpshooter Homalodisca vitripennis (Hemipterae: Cicadellidae) in French Polynesia and has successfully controlled about 95 percent of the pest density. Parasitiods comprise a diverse range of insects that lay their eggs on or in the body of an insect host, which is then used as a food for developing larvae. Parasitic wasps take much longer than predators to consume their victims, for if the larvae were to eat too fast they would run out of food before they became adults. Such parasites are very useful in the organic garden, for they are very efficient hunters, always at work searching for pest invaders. As adults, they require high–energy fuel as they fly from place to place, and feed upon nectar, pollen and sap, therefore planting plenty of flowering plants, particularly buckwheat, umbellifers, and composites will encourage their presence.

Four of the most important groups are:

- Ichneumonid wasps: (5–10 mm) Prey mainly on caterpillars of butterflies and moths.

- Braconid wasps: Tiny wasps (up to 5 mm) attack caterpillars and a wide range of other insects including greenfly. It is a common parasite of the cabbage white caterpillar, seen as clusters of sulphur yellow cocoons bursting from collapsed caterpillar skin.

- Chalcid wasps: Among the smallest of insects (<3 mm). It parasitizes eggs/larvae of greenfly, whitefly, cabbage caterpillars, scale insects, and strawberry tortrix moth.

- Tachinid flies: Parasitize a wide range of insects including caterpillars, adult and larval beetles, true bugs, and others.

Parasitic Nematodes

Nine families of nematodes (Allantone-matidae, Diplogasteridae, Heterorhabditidae, Mermithidae, Neotylenchidae, Rhabditidae, Sphaerulariidae, Steinernematidae, and Tetradonematidae) include species that attack insects and kill or sterilize them, or alter their development. In addition to insects, nematodes can parasitize spiders, leeches, [[annelid[[s, crustaceans and mollusks. An excellent example of a situation in which a nematode may replace chemicals for control of an insect is the black vine weevil, Otiorhynchus sulcatus, in cranberries. Uses of chemical insecticides on cranberry either are restricted or have not provided adequate control of black vine weevil larvae. Heterorhabditis bacteriophora NC strain was applied, and it provided more than 70 percent control soon after treatment and was still providing that same level of control a year later.

Many nematode–based products are currently available. They are formulated from various species of Steinernema and Heterorhabditis. Some of the products found in various countries are ORTHO Bio–Safe, BioVector, Sanoplant, Boden-Ntitzlinge, Helix, Otinem, Nemasys, and so forth. A fairly recent development in the control of slugs is the introduction of "Nemaslug," a microscopic nematode (Phasmarhabditis hermaphrodita) that will seek out and parasitize slugs, reproducing inside them and killing them. The nematode is applied by watering onto moist soil, and gives protection for up to six weeks in optimum conditions, though is mainly effective with small and young slugs under the soil surface.

Plants to Regulate Insect Pests

Choosing a diverse range of plants for the garden can help to regulate pests in a variety of ways, including:

- Masking the crop plants from pests, depending on the proximity of the companion or intercrop.

- Producing olfactory inhibitors, odors that confuse and deter pests.

- Acting as trap plants by providing an alluring food that entices pests away from crops.

- Serving as nursery plants, providing breeding grounds for beneficial insects.

- Providing an alternative habitat, usually in a form of a shelterbelt, hedgerow, or beetle bank, where beneficial insects can live and reproduce. Nectar–rich plants that bloom for long periods are especially good, as many beneficials are nectivorous during the adult stage, but parasitic or predatory as larvae. A good example of this is the soldier beetle, which is frequently found on flowers as an adult, but whose larvae eat aphids, caterpillars, grasshopper eggs, and other beetles.

The following are plants often used in vegetable gardens to deter insects:

Plant	Pests
Basil	Repels flies and mosquitoes.
Catnip	Deters flea beetle.
Garlic	Deters Japanese beetle.
Horseradish	Deters potato bugs.
Marigold	The workhorse of pest deterrents. Discourages Mexican bean beetles, nematodes and others.
Mint	Deters white cabbage moth, ants.
Nasturtium	Deters aphids, squash bugs and striped pumpkin beetles.
Pot Marigold	Deters asparagus beetles, tomato worm, and general garden pests.
Peppermint	Repels the white cabbage butterfly.
Rosemary	Deters cabbage moth, bean beetles and carrot fly.
Sage	Deters cabbage moth and carrot fly.
Southernwood	Deters cabbage moth.
Summer Savory	Deters bean beetles.
Tansy	Deters flying insects, Japanese beetles, striped cucumber beetles, squash bugs and ants.
Thyme	Deters cabbage worm.
Wormwood	Deters animals from garden.

Pathogens to be used as Biopesticides

Various bacterial species are widely used in controlling the pests as well as weeds. The best–known bacterial biological control which can be introduced in order to control butterfly caterpillars is Bacillus thuringiensis, popularly called Bt. This is available in sachets of dried spores, which are mixed with water and sprayed onto vulnerable plants such as brassicas and fruit trees. After ingestion of the bacterial preparation, the endotoxin liberated and activated in the midgut will kill the caterpillars, but leave other insects unharmed. There are strains of Bt that are effective against other insect larvae. Bt. israelensis is effective against mosquito larvae and some midges.

Viruses most frequently considered for the control of insects (usually sawflies and Lepidoptera) are the occluded viruses, namely NPV, cytoplasmic polyhedrosis (CPV), granulosis (GV), and entomopox viruses (EPN). They do not infect vertebrates, non–arthropod invertebrates, microorganisms, and plants. The commercial use of virus insecticides has been limited by their high specificity and slow action.

Fungi are pathogenic agents to various organisms including the pests and the weeds. This feature is intensively used in biocontrol. The entomopathogenic fungi, like Metarhizium anisopliae, Beauveria bassiana, and so forth cause death to the host by the secretion of toxins. A biological control being developed for use in the treatment of plant disease is the fungus Trichoderma viride. This has been used against Dutch Elm disease, and to treat the spread of fungal and bacterial growth on tree wounds. It may also have potential as a means of combating silver leaf disease.

Significance of Biological Control

Biological control proves to be very successful economically, and even when the method has been less successful, it still produces a benefit–to–cost ratio of 11:1. The benefit–to–cost ratios for several successful biological controls have been found to range from 1:1 to 250:1. Further, net economic advantage for biological control without scouting vs. conventional insecticide control ranged from $ 7.43 to $ 0.12 per hectare in some places. It means that even if the yield produce under biological control be below that for insecticidal control by as much as 29.3 kilos per hectare, the biological control would not lose its economic advantage.

Biological control agents are non–polluting and thus environmentally safe and acceptable. Usually they are species specific to targeted pest and weeds. The biological control discourages the use of environmentally and ecologically unsuitable chemicals, so it always leads to the establishment of natural balance. The problems of increased resistance in the pest will not arise, as both biological control agents and the pests are in complex race of evolutionary dynamism. Because of chemical resistance developed by the Colorado potato beetle (CPB), its control has been achieved by the use of bugs and beetles (Hein).

Negative Results of Biological Control

Biological control tends to be naturally self–regulating, but as ecosystems are so complex, it is difficult to predict all the consequences of introducing a biological controlling agent. In some cases, biological pest control can have unforeseen negative results, that could outweigh all benefits. For example, when the mongoose was introduced to Hawaii in order to control the rat population, it predated on the endemic birds of Hawaii, especially their eggs, more often than it ate the rats. Similarly, the introduction of the cane toad to Australia 50 years ago to eradicate a beetle that was destroying sugar beet has been spreading as a pest throughout eastern and northern Australia at a rate of 35 km/22 mi a year. Since the cane toad is poisonous, it has few Australian predators to control its population

Companion Planting

Companion planting is the art of growing plants in proximity to each other because of their ability to enhance or complement each other. The practice has been used by farmers and gardeners in both industrialized and developing countries for many reasons. Many plants produce natural substances in their roots, flowers, leaves that can attract or repel insects depending on your garden needs. The benefits of companion planting include pest control, nitrogen fixation, providing support of one plant by another, enhancing nutrient uptake, and water conservation among other benefits. Hence companion planting can lead to increased yield, less reliance on pesticide, and increased biodiversity, helping to bring a balanced eco-system to your garden and allowing nature to do its job. Nature integrates a diversity of plants, insects, animals, and other organisms into every ecosystem so there is no waste. The death of one organism can create food for another, meaning revolving benefits all around. Companion planting is considered to be a holistic concept due to the many interrelated levels it cooperates with the ecology.

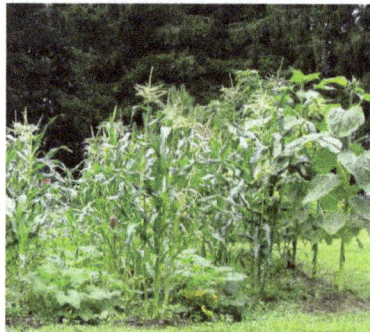

Companion planting involving corn, beans, squash and sunflower. The corn acts as a natural beanpole, while the beans contribute nitrogen to the soil for the other crops to use. Squash vines can discourage raccoons, while its leaves provide shading to control weeds and reduce water loss by evaporation. Sunflower attracts beneficial insects such as pollinators.

Companion planting is not a new concept; many of the modern practices were used many centuries ago in ancient gardens. The practice is considered to be a form of poly-culture, which means the raising of multiple crops in the same space, reflecting the diversity of natural ecosystems, and avoiding large stands of single crops or monoculture.

By using companion planting, many gardeners find that they can discourage harmful pests without losing the beneficial allies. There are many varieties of herbs and flowers that can be used for companion plants, but the suggested way succeed is to keep experimenting to find what works for you. However, you may want to include plants that are native to your area so the insects you want to attract already know what to look for. Plants with open cup shaped flowers have been reported to be the most popular with beneficial insects. Some selected vegetable plants and their suitable and unsuitable or poorly suited companions are listed in Table. Suitable companions provide benefits such as nitrogen fixation, production of invigorating exudates, repelling or trapping of insect and other pests, and weed suppression among other benefits. The unsuitable or poorly suited companions are those that are detrimental to the crop by either inhibiting its growth or encouraging pest or disease proliferation.

Benefits of Companion Planting

Aesthetics

One benefit derived from companion planting includes combining beauty and purpose, giving you an enjoyable, healthy environment. You can turn your backyard into a garden paradise. You can make gardening fun and appealing, and only limited to your imagination. There are countless ways you can incorporate companion plants in your backyard garden, orchard, and flower beds.

Application in Trap Cropping

Trap cropping is the planting of a trap crop to protect the main crop from a certain pest or several pests. The trap crop can be from the same or different family group, than that of the main crop, as long as it is more attractive to the pest. A crop may be selected because it is more attractive to pests and serves to distract them from the main crop, for example planting collards to draw the diamond back moth pest away from cabbage.

Improving Soil Fertility

Soil fertility can be improved by incorporating plant that can fix atmospheric nitrogen. Legume plants such as beans, peas and clover, have root nodules that harbor Rhizobium bacteria that help to fix nitrogen through a symbiotic relationship with plants. The symbiont plants will have nitrogen for their own use and for the benefit of the neighboring non nitrogen-fixing plants.

Biological Pest Control

Companion planting can enhance biological pest suppression through allelopathy, which is a biological phenomenon where an organism produces one or more biochemicals that influence the growth, survival, and reproduction of other organisms. Some plants exude chemicals from roots or aerial parts that suppress or repel pests and protect neighboring plants. For example, African marigold is reported to release thiopene which act as a nematode repellent; hence it can act as a good companion for a number of garden crops.

Beneficial Spatial Interactions

Companion planting can provide beneficial spatial interactions where tallgrowing, sun-loving plants may share space with lower-growing, shade-tolerant species, resulting in higher total yields from the land. Spatial interaction can also yield pest control benefits by acting as a deterrent for pest. Such a benefit for example, has been reported when members of the gourd family are interplanted with sweet corn and beans. It is believed that such intercrop disorient the adult squash vine borer and protect the vining crop from this damaging pest. In turn, the presence of the prickly vines is said to discourage raccoons from ravaging the sweet corn and beans, while the bean crop provides nitrogen for the plants.

Nurse Cropping

Nurse cropping involves planting a crop in the same field with another crop, to provide benefits such as minimizing the growth of weeds. For instance, tall or densecanopied plants may protect more vulnerable species through shading or by providing a windbreak. Nurse cropping can be viewed as another version of spatial interaction where by you can quickly grow a crop in an unused area next to another crop that has a longer growing cycle. An example is growing broccoli and lettuce. By the time broccoli gets large enough, the lettuce below will benefit from the shading of the large waxy leaves, extending the growing season of lettuce and preventing bolting.

Providing Security Through Diversification

By simply mixing plants in your garden you are giving yourself some security in case of a crop failure. If pests or adverse weather conditions reduce or destroy a single crop or cultivar, others remain to produce some level of yield.

Table: Vegetable crops, their suitable and unsuitable or poorly suited companions.

Crop	Suitable companions	Unsuitable or poorly suited companions
Asparagus	Tomato, parsley, basil	Onion, garlic, potato
Basil	Tomato	Rue
Beans	Carrot, cabbage, cauliflower, corn, cucumber, rosemary	Leek, garlic, shallots, chives

Beans, bush	Irish potato, cucumber, corn, strawberry, celery, summer savory	Onion, garlic
Beans, pole	Corn, summer savory, radish	Onion, beets, kohlrabi, sunflower
Beet	Cabbage, onions, kohlrabi	Pole beans, field mustard
Cabbage family	Aromatic herbs, celery, beets, onion family, chamomile, spinach, chard	Dill, strawberries, pole beans, tomato
Carrots	Pea, lettuce, onion, rosemary, tomato	Dill, parsnip, radish
Celery	Onion & cabbage families, tomato, bush beans, nasturtium	Parsnip, potato
Corn	Irish potato, beans, english pea, pumpkin, cucumber, squash	Tomato
Cucumber	Beans, corn, pea, sunflowers, radish	Irish potato, aromatic herbs
Eggplant	Beans, marigold	
Lettuce	Carrot, radish, strawberry, cucumber	
Onion family	Beets, carrot, lettuce, cabbage family, summer savory	Beans, english peas
Parsley	Tomato, asparagus	
Pea, english	Carrots, radish, turnip, cucumber, corn, beans	Onion family, gladiolus, irish potato
Potato, irish	Beans, corn, cabbage family, marigolds, horseradish	Pumpkin, squash, tomato, cucumber, sunflower
Pumpkins	Corn, marigold	Irish potato
Radish	English pea, nasturtium, lettuce, cucumber	Hyssop
Spinach	Strawberry, faba bean	
Squash	Nasturtium, corn, marigold	Irish potato
Tomato	Onion family, nasturtium, marigold, asparagus, carrot, parsley, cucumber	Irish potato, fennel, cabbage family
Turnip	English pea	Irish potato

Permaculture

Permaculture is a sustainable design system that applies ecological principles that are found in nature to the development of human settlements, allowing humanity to live in harmony with the natural world. While permaculture can be applied to almost any area of living, including local economies, energy systems, water supplies, and housing systems, permaculture has become most well known for its applications in sustainable food production.

Permaculture strives to "work smarter, not harder," banish waste in all its forms (such as pollution, water waste, and energy waste), and increase natural productivity and efficiency over time through the application of sustainable design systems that work with nature within its natural limitations. Permaculture is especially useful in a world where there are constrained energy and natural resources.

In permaculture, there is particular emphasis on the use of perennial crops such as fruit trees, nut trees, and shrubs that all function together in a designed system that mimics how plants in a natural ecosystem would function.

The Permaculture Ethics

All permaculture design and practice is informed by three permaculture ethics:

Permaculture ethics are the foundation of permaculture design. Permaculture designs will take more initial work and an understanding of – or a willingness to learn about – the complexity of natural systems. Because of this, we will only feel motivated to design with permaculture strategies if we value the ethical standards on which the permaculture approach was created.

The permaculture ethics are simply: care for Earth, care for people, and reinvesting abundance.

Care for Earth

We are only as healthy as our planet. Caring for the forests, the waterways, and the diverse life forms of our magnificent planet benefits us. On your land, actively seek ways to regenerate fertility and biodiversity rather than simply sustaining current levels.

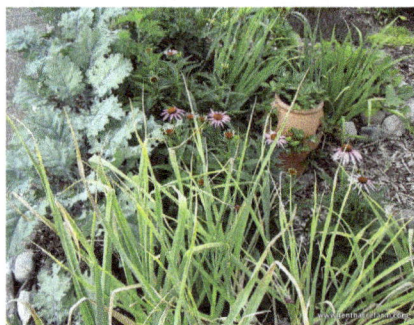

Care for People

Caring for people includes caring for ourselves and our own household. When we 'take responsibility for our own existence', we inevitably begin producing more and consuming less. It is this step away from consumerism that also helps us avoid products and companies that exploit people.

In modern times, it has become admirable to favor the opposite of taking responsibility for ourselves: Committing our lives to helping others, which in turn leaves little room to care for ourselves, little time to achieve any level of self-sufficiency, and little energy for reducing our own level of consumption. This unfortunately can have a net zero effect.

Reinvesting Abundance

When we care for the Earth, nature responds with abundance—more biodiversity, more plants, more animals, healthier water, healthier air, and so on. We can reinvest useful flows—such as rainwater or compost—back into the system to create a self-maintaining ecosystem that requires fewer inputs from off-site sources.

This is the pinnacle of land conservation: Honoring and encouraging the abundance of the land we inhabit, rather than viewing our resources as scarce with a focus on importing materials.

When we care for ourselves and act as responsible consumers, life becomes abundant. We have access to an abundant supply of healthy, homegrown food. We are financially more resilient. Ultimately, caring for our own existence provides abundance that can be reinvested into our community—through sharing food, skills, or financial assistance. This is abundance.

Permaculture Principles

Permaculture is rooted in the fact that no single problem or solution stands on its own. In recognition of this balance, it embraces four basic principles.

Working with Nature Rather than Against it

It seems that humans often try to make things more difficult than they need to be. It has been proven that legumes, compost, and organic matter all help to feed the soil free of charge, but we continue to research and pay for artificial fertilizers. We know that growing a variety of plants is one of the keys to healthy living. It allows us to improve our soil, protect our environment, and meet our nutritional needs. It also offers us a renewable source of things like medicines, building supplies, fuel, clothing, etc. And yet we often put all of our energy into clearing away large areas of natural diversity to make room for only one or two food crops. Our ancestors collected and used indigenous seeds, which are better adapted to the areas in which they grow, they are often pest and drought resistant, and best of all they are free— yet we still spend our money on foreign hybrids. We have also been shown that nature has ways of controlling pests and diseases, but we still prefer to purchase chemicals to do these jobs. Why? It seems that we often work overtime to struggle against nature. Each year during the rainy season, thousands of seedlings spring up from our soil. These plants, if allowed to mature, could be providing Malawi and the rest of the world with nutritious food, firewood, building supplies, and everything that we need for our survival. But instead of simply allowing nature to do all of this work for us, we waste our energy trying to get rid of these things. Through practices like burning, clear-cutting, monocropping, over-grazing and over-sweeping, we eliminate all that nature has given to us as a gift. Nature is a miracle that is taking place every day right outside of our doorstep. If we simply start working with this miracle, rather than against it, our rewards could be unlimited. The world is over-flowing with potential.

Luwayo Biswick teaching American students about bore-hole design.

Thoughtful Observation Rather than Thoughtless Labor

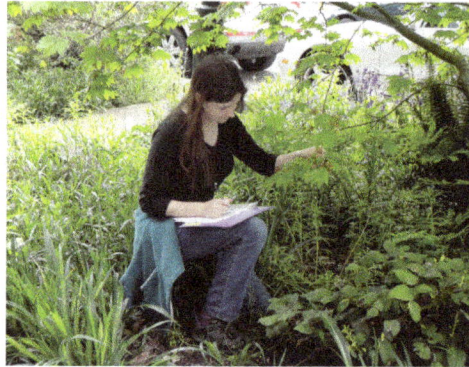

Observing and Learning from nature.

We can learn a lot from nature if we give it the chance to teach us. Since the beginning of time, generations of people have known this and used it to their advantage. The earliest sailors used the stars for navigation, the sky to predict the weather, and the behavior of animals to locate sources of food. Traditional healers from every part of the world have used observation of nature to show them plants that can heal us, and those which could harm us. If we take the time for these observations, we can see how every part of nature, from the smallest microscopic organisms to the largest plants and animals, all perform a specific function. Each of their individual roles is vital for the success of the whole. Permaculture advises us to make the most of these observations before we put our energy into something that might be working against nature, rather than with it.

Each Element should Perform many Functions, Rather than One

If we have made use of our thoughtful observations, we will be able to see that each element in nature, even though it may have a specific function, is probably carrying out many other functions as well. If we pay attention to these observations, we can utilize one thing for many reasons and drastically improve our lives and our environments. If we plant a mango tree, for instance, we know that it will eventually give us the nice juicy fruit that we love to eat. But if we look at this mango tree through the eyes of Permaculture, we learn to recognize all of the other functions that it will be performing. This same mango tree that we are planting for fruit will also give us: shade, medicine, firewood, and protection from the wind. Its leaves will break the force of the falling rain and give us organic matter for our compost piles. Its roots will help to hold the soil in place and bring up minerals from deep below the soil's surface. Its branches will provide shelter for all sorts of birds, insects, and other wildlife. And then, on top of all of this, it will yield the delicious food that will give us nutrients that our bodies need to stay healthy and strong. If we take into consideration all of the mango trees functions, we can then decide where to place it or what to put near it to make use of the most of its uses. When we start to look at nature in this manner, we can see that we have all of the resources that are necessary to sustainably meet our needs for long into the future.

Everything is Connected to Everything Else

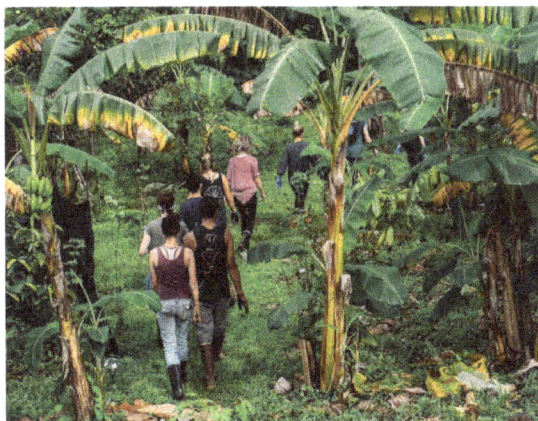
Howard's integrated food forest.

As human beings, our connection with nature is an essential component of Permaculture. For some reason, we often think of ourselves as somehow separate or above the laws that govern all of nature. This type of thinking has isolated us from the very roots of our existence. As soon as we place ourselves back into the cycle of nature, it becomes clear that whatever we do to this cycle will have an impact on our survival. If we strengthen and nurture it, it will yield rich rewards for our future, but if we continue to take away from this cycle and weaken it, the outcome will have devastating results. Life truly is a web, of which humans are only one strand. If the strands of this web are weakened or broken it will no longer be able to support itself. On the other hand, if we do everything in our power to replace, mend, and reinforce the strands of this web it will mean a better future for the entire planet. Nature comes equipped with checks and balances. It has methods of controlling diseases, regulating its population, and healing itself. If we upset this balance, we often end up creating many unforeseen problems for ourselves. The deliberate extermination of one insect that may be causing damage to our crops can have disastrous results. It may be that the so-called "pest" we have decided to eliminate is responsible for keeping the population of even more harmful insects under control. Or perhaps, that so-called "pest" is an intricate strand in nature's web that other animals depend upon. This was the case that we tragically learned in America through the use of an insecticide known as DDT. The insects that were eliminated through the use of this chemical were part of a larger food chain. The contamination of this food chain eventually resulted in bringing the national symbol of the United States, the American Bald Eagle, close to extinction.

Permaculture Designs

Most people are introduced to permaculture in the garden. This holistic approach to gardening offers exciting, effective, efficient ways to produce food in small spaces without chemicals and without waste.

Instead of using the conventional rows and crops most people picture when they imagine a garden, permaculture garden designs utilize plants and planting methods that are better suited to the specific area and climate rather than striving to grow something that matches poorly with the environment.

This approach may mean that not all of the fruits and vegetables shoppers are accustomed to eating will be available in their backyard garden. But it does equate with a more low-maintenance, positive-impact abundance of a wide variety of crops.

That means growing local, fresh, organic foods that improve the environment as opposed to chemically derived, internationally imported products. Most of which come from monoculture and factory farms that create pollutants and encourage deforestation.

Permaculture designs can be implemented at a mass market scale. But the real crux of the movement is the idea that society as a whole would benefit from moving back towards home garden production, both in rural and especially urban settings.

The following examples of permaculture gardening techniques can help to establish sustainable soils that will make your fruits and vegetables grow as sustainably and as nutritious as possible.

No-Dig Bed.

No Dig Gardening

No-dig gardening is a widely applied permaculture technique designed to preserve the soil life that converts organic matter into plant food.

When we dig or till, we kill beneficial bacteria, organisms, and creatures that keep the soils in good order. Initially, the decomposition of their bodies will feed the plants in our garden.

But ultimately, when there's no soil life left to decompose, the result will be nutrient-deficient soils. Therefore, we should try to encourage soil life rather than deplete it.

No-dig gardening beds are built atop existing soil, which has several benefits. Beyond saving soil organisms, the topsoil we're building the bed on and the soil and mulch we're using provide a double dose of quality growing medium.

For those in humid climates, no-dig beds help to make sure that soils drain well. The use of deep mulches ensures soils stay sufficiently moist, cutting down on maintenance and water use. Popular no-dig methods include raised beds, sheet mulched beds (my favorite), and hugelkultur.

African Keyhole Bed.

Keyhole Garden Designs

Permaculture gardening also emphasizes efficient design, and keyhole beds are a great example. In conventional agriculture, we only plant about 50% of the land we're cultivating. The raised rows are planted, while the troughs are more or less left to weeds.

With keyhole design beds, we're able to maximize ROI from the space we've got for our garden. Keyholes also encourage biodiversity, using mixed planting rather than rows of singular crops (which are more susceptible to diseases and pests).

With a keyhole garden, more land goes to cultivation than dead space. This is accomplished by bending a row three to four feet wide around a central point. You'll want to leave a small access path to that point so that the bed resembles a keyhole.

These garden beds can either be harvested from the center, or that spot can be where your compost is concentrated and leached into the raised garden. Because everything is condensed into one spot, it takes less effort to harvest, and we're using more of our garden for growing plants.

Worm Composting

Worm Composting

Worm Composting or vermiculture is when worms create compost, and vermiculture buckets are a very efficient way of doing this. Rather than having a single compost bin where we put all of our food scraps, we space many small bins around the garden. Kitchen scraps are distributed into the different buckets, into which composting worms have been introduced.

The worms break down the kitchen scraps into worm castings that are much more fertile than typical compost. This permaculture gardening technique works great with both raised beds and keyhole designs. Simply drill a bunch of nickel-sized holes into the bottom half of a 5-gallon bucket. Then bury the bucket in the garden, with the bottom half beneath the soil surface.

Then fill the bucket with a bed of shredded paper and cardboard topped off with a layer of soil, manure and dried grass. Once composting worms are added, they'll handle the rest. The nutrients of the vermicompost drain directly into the garden and feed the plants. When the bucket is full, the castings can be spread over the bed.

Comfrey

Chop-and-Drop Organic Mulch

Mulching is another key difference between permaculture gardening and conventional gardening, which typically removes all organic matter. Mulch, especially organic mulch, offers a wide range of benefits. It moderates soil temperature by keeping the sun off of it and insulating it from the cold. It prevents evaporation so that the soil stays moist. It stops heavy rains and wind from eroding the topsoil. And, as it breaks down, it reinvigorates the soil with more organic matter to feed on.

Chop-and-drop mulch is a technique that doubles up on the benefits. By growing nitrogen-fixing legumes and dynamic accumulators (i.e. plants with deep taproots that pull minerals up from deep in the earth), top-quality mulch material is produced right there on-site.

Nitrogen-fixing legumes provide natural fertilizer—nitrogen—while the decomposition of dynamic accumulators re-mineralizes the soil. Because the plants are in the garden, leaves and branches are simply chopped and dropped right onto the beds. These plants will grow back again and again, providing more and more organic matter to constantly improve the soil.

Carrots and Onions

Companion Gardening

Biodiversity in the garden is beneficial because the nutrient needs of crops vary. Pests aren't provided with rows upon rows of their favorite food. And harvests aren't centered on the success of one crop.

Companion gardening ups the ante with pairings in which the individual plants provide services for the other plants in the group. Having chop-and-drop organic mulch plants is an example of this, but there are many other aspects of companion planting.

Good plant combinations can help with controlling pests, attracting beneficial insects, filling vertical spaces, and providing fertilization. Most culinary herbs and many flowers, such as Nasturtium and Marigold, are fantastic repellents for garden pests. Flowering plants are very good for attracting bees and other pollinating insects.

The trinity of corn, bean, and squash are a classic combination of efficiently using vertical space in your garden. The corn grows high, beans use their stalks as poles, and squash winds along the ground. Corn, beans, and squash also have beneficial relationships at the root level, from which each plant's growth is enhanced.

Ruby Chard

Rotational Cropping

Perennial plants are a huge part of permaculture design because they supply food without needing to be replanted year after year. But that isn't to say that permaculture gardening plans never include annual plants.

Annuals do have a place in permaculture, but the overall approach is more measured. Growing annuals sustainably requires rotating crops, which means changing the type of plants we are growing each time we cultivate a particular bed.

When the same plant (or plant family) is used in the same soil time and time again, that soil becomes depleted of whatever nutrients that plant likes. And diseases and pests that like that particular crop linger in the soil, waiting for the next round of feasting.

By using rotational cropping techniques we can create simple sequences of planting that will revitalize the soil and keep pests and diseases at bay.

A typical sequence starts with soil-enhancing beans and peas before hungry cruciferous vegetables. Then you switch to nightshades like tomatoes, peppers, and eggplants, with mineral-mining root vegetables as the last leg. Then, the whole cycle starts over.

Red Clover

Green Manure

Permaculture gardening centers on three core ethics: Earth care, people care, and fair share. In terms of gardening, "fair share" means that we can't constantly take from our gardens without giving back.

Conventional agricultural has done this in the form of chemical fertilizers, which is neither necessary nor sustainable.

Using fertilizer as opposed to organic matter to feed plants doesn't help with feeding soil life and building new soils. The end result of this process is completely sterile earth that has nothing more to give in terms of nutrients.

Green manure is another way to give back to the garden. This process involves growing soil-amending groundcovers with the sole purpose of cutting them down and giving them right back to the garden as organic matter.

Natural systems work this way: Things grow, die, and go back to the earth so that the next generation can grow from them. If we only ever harvest from our annual gardens, the soil never gets back the energy lost in producing those crops.

It's only fair to our soil to occasionally grow something for it. Many gardeners grow green manure as part of the crop rotation (often after the root vegetables).

Keyline Design

Keyline methods enable the rapid development of deep biologically fertile soil by converting subsoil into living topsoil. Keyline pattern cultivation enables the rapid flood irrigation of undulating land without terracing. Incidental results are the healing of soil erosion, bio-adsorption of salinity and the long term storage of atmospheric carbon in the soil as humus.

Keyline farm planning is a management tool that uses natural landscape contours and farming techniques to slow, sink, spread and store rainwater as well as build soil fertility. With a detailed contour map of your farm, keyline planning can help determine the optimal placement for farm elements such as: irrigation ponds, cropping & orchard rows, structures, roads/tracks, fences, livestock rotation, subsoil rip lines, and more.

Keyline plowing uses a subsoil plow method with a very flat plow shank (about 8%) to slice through the soil and create channels below the surface. These channels help break up soil compaction, create a place for new roots to grow with less effort and direct water more easily.

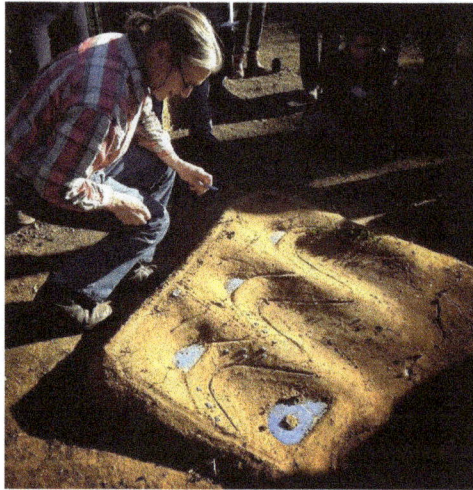

David Holmgren demonstrating how keyline works.

Keyline is a 'Land planning system' with the main focus being on the control of water resources. Observing where it is, capturing it, holding it, cycling it through as many systems as possible before allowing it to leave.

Keyline Plan

Key components:

- Rapid development of biologically active, fertile soil within a systematically designed landscape. During an average three-year conversion phase, four to six inches of new topsoil are typically formed each year. This new topsoil stores large quantities of water in the landscape.

- Design for the harvest, storage and distribution of water on the landscape forms the foundation of the Keyline Plan.

- Run-off water is stored in dams. This water is later released for rapid, gravity-powered flood-irrigation.

- Roads, forests buildings and fencing follow primary water layout and fit together within the lay of the land.

- The Keyline landscape is a permanent landscape in which every infrastructure component helps ensure the maintenance and renewal of the topsoil within it.

New Topsoil can be Created Quickly

Factors that determine soil fertility:

- The mineralogical and structural framework

- The prevailing climate

- The soil's biotic associations.

Soil has a life and environment of its own. The biotic association can be modified through modification of the soil microclimate.

Soil life responds dramatically to ideal air, moisture, food and temperature conditions. These conditions are simple to create with grazing, subsoiling and dependable rainfall or irrigation. Life begets Life. Plants, their roots and attendant exudates are the solar harvesters and the raw food of soil life. Grazing animals are 'biological accelerators' they are the most effective tool we can use to speed mineral cycling, and graziers affect enough land to make a large impact.

Keyline Planning is based on permanence, beginning with the two most permanent features of the landscape:

- Climate which has moulded and created the topography of the dominant climatic factors, temperature, wind, annual distribution of humidity, rainfall. Water is the easiest to work with ("control") and gain benefit from.

- Existing Land Shape and form (Topography) including underlying geology.

Combining Holistic Management Land Planning with Keyline Planning:

- Form a Holistic Goal, including detailed land/ecosystem process description in the Future Resource Base.

- Get Topographical Maps. Analyze landscape using Keyline insights. Identify Keypoints, Keylines, ideal water storage areas, water diversion lines, possible irrigable areas, road layouts, tree lines, etc.

- Gather all pertinent information, study and prepare maps and overlays. Take a year or two.

- Brainstorm many possible layouts for the planned developments.

- Create the ideal plan based on the best ideas.

- Develop the plan gradually through Holistic Management Financial Planning so that each investment makes rather than costs money.

Holistic Management Planned Grazing and Keyline Soilbuilding go hand in hand. The growing season grazing plan gives you a structured, holistic framework to plan the use of tools (grazing animal impact, subsoiler plow) in the soilbuilding project.

Water Control is Paramount

Water and rainfall determine land development. We have to get water right to get everything else right - design follows water.

New, "artificial" water lines diversions, dam walls, channels - become permanent land features. Other infrastructure components follow.

Direct rainfall and irrigation water are spread evenly on the land by a unique cultivation pattern, which is an artificial water line Keyline Cultivation.

Natural Water Lines

Water flowing over land has a pattern of flow and predictable path lines of movement.

- The contour line the edge of a lake is a true contour line. Flow is perpendicular to the contour, forming shallow S-curves from the ridge to the valley.

- Water drainage lines- streams.

- Water divide lines – "watersheds," main ridge crests.

Artificial Water Lines

Human earthworks that influence flow of water and store water.

Diversions, irrigation channels, dams, Keyline Cultivation pattern, swales. (Also, drainage ditches, which are not central to Keyline.)

Artificial water lines in Keyline are designed for the most efficient water resource development. Proper design of farms and cities must fit with the existing design in the natural landscape.

Geography of Landscape

Three water lines, three land shapes & one special pattern.

Contour Water Line

The shore of a lake. A level line running across the landscape, a set vertical distance from the next contour line. Water will always run perpendicular to the contour.

Water Drainage Line

The centre of watercourses: streams, rivers drainage lines of the land. Dendritic (branching) patterns.

Water Divide Line

Crests of Main and Primary Ridges

Vegetation slows the movement of water over and through the land. Vegetation, its variety and its absence, and soil organisms stabilize soil and land shapes.

In a stabilized landscape, there are three land shapes we consider in relation to Keyline development:

1. The main ridge

2. The primary valley

3. The primary ridge.

The main ridge is the first land shape. It begins at the convergence of two water drainage lines. Look around it is the horizon. The crest of a ridge is synonymous with a water divide line. The crest of a ridge is usually less steep than the sides of the ridge.

Main ridges are a reverse image of the dendritic branching of water drainage lines (streams/rivers). You could follow main ridges around the world, except where they go in circles around lakes. The interplay of main ridges and water drainage lines are the anatomy of the landscape.

Primary Valleys form in (erode into) the sides of Main Ridges. Primary Valleys are divided by Primary Ridges. A primary valley has a primary ridge on either side, so there is always one more primary ridge than primary valleys in a main ridge system. Primary Valleys are the first place water flows in a rainstorm. Primary valleys are the smallest of the three land shapes. They are the only true "valley" shapes in the landscape. (Big valleys are actually watersheds.)

The centreline of a primary valley is usually less steep than the sides of the valley. Where a primary valley intrudes far into a main ridge.

In the figure below a saddle is described as roads that usually cross over main ridge crests across saddles. Next to a saddle is a hill.

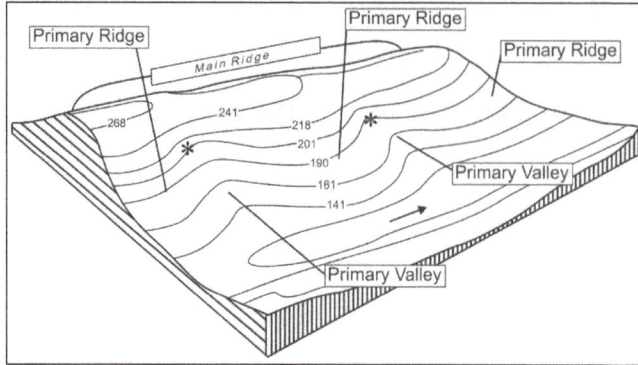

Lakes and ponds are located in depressions in the landscape.

Walk up from the end of a main ridge, (above the confluence of two streams) and it eventually runs into another ridge you can go left or right on a main ridge. This pattern repeats endlessly. It almost seems designed to shed water. It is primarily the result of the underlying geological skeleton, the urge of water to get back to the sea (water flows downhill) and the moderating influence of vegetation and soil life.

Fragments Between:

- Tidal Areas

- Flood Plains

Keypoint

Every primary valley has a keypoint. It is the point at which the primary valley gets suddenly steeper. The steepest slopes in the landscape usually occur in the centre of the valley above the Keypoint between the Keypoint and the top of the main ridge.

Identifying the Keypoint and attendant Keyline, is the starting point for Keyline design.

The Keyline is a contour line carried in both directions from the Keypoint, in the valley shape, but not extending out onto the ridges. Below the Keyline, the accompanying (next-door) primary ridge centre is steeper than the primary valley centre.

Above the Keyline, the primary valley centre is steeper than the primary ridge centre.

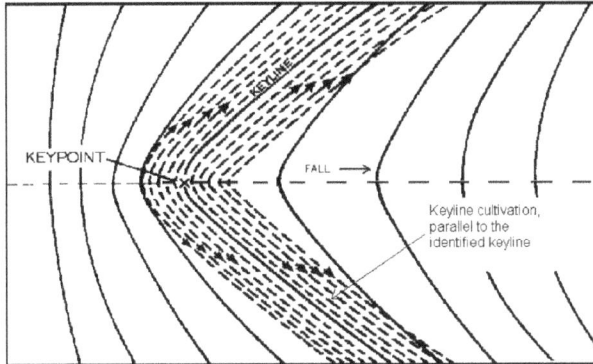

Cultivate parallel to the Keyline both above and below the keypoint in the valleys. Cultivate parallel and upward from any selected contour line on the ridges. When there is no Keyline to work from (lower in the valleys, or on ridges) use contour guidelines to cultivate parallel to (upward from on ridges, downward from in valleys.) This is Keyline pattern cultivation. Water will drift from the valley shapes toward the ridges.

Main Ridge

Main ridges occupy the most land in the landscape. They are not level, but slope. This creates a rising relationship in the Keypoints of adjacent primary valleys.

Contour Maps are Basic to Understanding Keyline

Contour maps show the above land features clearly. Contours are level lines, a set vertical distance from each other. Close lines indicate steep land, more widely spaced lines indicate less steep land.

With a good contour map and the farmer to provide details, we can design a landscape which will include:

- Water diversion, storage, irrigation channels, irrigable land and water control structures,
- Catchment size,
- Slope Indices,
- Size of dam walls,
- Areas to leave, plant, cut trees,
- Sites for farm buildings,
- Location of subdivision fences, stock watering points, paddock layout.

Contour maps from Government sources, especially of rural landscapes, only provide contour intervals of 10-20+m. Surveyor-produced contour maps are more expensive but are very accurate and provide contour intervals of between 100mm (very flat landscapes) and 1-2m for more undulating or steep landscapes.

Overlaying contours onto an Aerial Photo provides an advanced base to design a landscape with. Using Geographic Information System (GIS) and Computer Aided Design (CAD) software can really enhance the design and development potential of a landscape and form a base from which to easily create a 'Bill of Quantities' for all aspects of the landscape and its development.

Categories of Water Available to a Farm

- Absorbed Rainfall – high quality, low price. Good soil holds great quantities of water. Developing topsoil is probably the most cost-effective way to enhance the water cycle and store water on the farm.

- Run off from rain falling on the farm. Rainfall has exceeded the field capacity of the soil, and runs off. Poor design will accelerate this.

- External Sources of Surface Water. Water flowing onto the farm.

- Ground Water- Pumped or spring fed.

Table: Change in the capacity of soil to store water (litres/ha) with changes in levels of soil organic carbon (OC) to 30 cm soil depth. Bulk density $1.2g/cm^2$.

Change in OC level	Change in OC (kg/m^{2})	Extra water $(litres/m^2)$	Extra water $(litres/ha)$	CO sequestered (t/ha)
1%	3.6kg	14.4	144,000	132
2%	7.2kg	28.8	288,000	264
3%	10.8kg	43.2	432,000	396
4%	14.4kg	57.6	576,000	528

Designing for the Environment

Understand the basic land shapes and design in accordance with enhancement of the water cycle by primarily slowing the movement of water over the land. (That is what life does, too.) Start as high as possible, by increasing the fertility and water holding capacity of the primary valleys and ridges. Maintain or develop productive or revegetation forests along the ridges and creek lines for landscape protection and to optimise nutrient or energy cycling, flows and utilisation.

Introduce artificial water lines: the diversion channel, the dam wall, the irrigation channel. Again incorporate productive or revegetation forest strips and plantings with these features.

Contours/Keylines

The Keyline is the contour line drawn through the keypoint.

Remember that keylines do not usually wrap continuously from one primary valley to the next. They have a rising relationship as one moves from one primary ridge to another.

On a contour map, the Keypoint is apparent, because the contour lines are closer together above it, and further apart below it. On a primary ridge, the centre of the ridge is typically flatter than the sides of the ridge, closer to the valley. (Contour lines are further apart in the centre, closer on the sides.) As the contour lines change direction and head into the valley, the lines will diverge if they are below the Keyline and converge if they are above the Keyline.

Water always flows perpendicular to the contour. This can be understood when we observe the heavier flow in the valleys, and the drier ridges.

Keyline Pattern Cultivation

Causes water to drift away from valley centres and toward ridge crests, where it is held until it soaks in. Rainfall and irrigation water are spread evenly over undulating land.

The simplest way to accomplish this, given an internal laser guidance system, or just a good feel for slope, is to plough slightly downhill from a given point in a valley centre out onto the accompanying ridge.

In primary valleys, we cultivate parallel to the Keyline above it and below it. Above it, it is often too steep for ploughing, but not always. The point at which we shift from valley pattern to ridge pattern cultivation, below the Keyline, is located where the valley floor becomes the ridge wall, or where the contour line shifts direction, in going from primary valley to primary ridge shape. This ploughing pattern will quickly become quite steep/angular, at which point a new contour line should be marked and ploughed parallel to and downward.

Anywhere lower in the primary valley, we cultivate parallel and below a contour guideline.

On primary ridges, we cultivate parallel and upwards from any contour guideline. It's good to stake a number of guidelines, i.e., not plough mindlessly too far from a guideline.

In practice, one would lay out the Keyline across the primary valley, then carry that contour line out onto and around both ridges, then cultivate upward from that in long plough passes. You would then plough downward from that line, restricting yourself to the valley shape. (The ridges would be ploughed parallel and upward from a lower contour guideline. In tighter valleys, there are tricks for simplifying difficult ploughing. However, the basic principles must be stuck by, or water will flow the wrong way, concentrating in the wrong places.

Keyline Scale of Permanence

1. Climate

2. Land Shape

3. Water

4. Roads

5. Trees

6. Buildings

7. Subdivision

8. Soil.

Water

Two Costs of Water

- Cost in money: Cost of improving soils, building dams and irrigation layout, irrigation operation.

- Cost in Water itself: It is expensive to always have water available. More cost effective to have water to bring you through dry times 100% drought-proofing would cost a fortune.

Stored Water, a 2nd Savings Account: Water in a dam can be traded for, say, a crop of pasture. A full dam, and dry fields in a drought is a sign of failure. Use water in dams for irrigation whenever necessary. Dams can and should be designed to be interlinked so as to be able to move water where it is needed during prolonged dry periods.

Farm Dams

Keyline dams always have a large pipe with baffle plates and a valve, through the bottom, for irrigation and control purposes.

Good sites for valley dams generally have:

- A flatter valley floor slope, backing water up further with less wall.

- A short wall site.

- Width of valley behind the dam wall.

- Suitable location for spillway.

- Suitable soils (will hold water).

- Suitable foundation material.

Highest site for a storage dam wall in a primary valley is below the Keypoint. This is called a Keypoint Dam. The Keyline is the top water level of the dam. Other types of dams include: Saddle Dams, Turkey Nests & Contour or Ridge Dams.

Water levels of dams can be connected by a diversion, falling at 1:400+. Or, water from the lockpipe of one dam can be carried by a diversion to the Keyline of another Keypoint Dam. Sometimes, a dam lower than at the Keypoint is desirable for a whole range of factors.

Other Types of Dams Include

Saddle Dams

Ring Dams

Contour Dams

Valley Dams

Water Channels

When developing the water resources of a farm, there are two primary water channels:

1. First, for diverting run off, stream flow or pumped water into a dam. Called a diversion or catchment drain and generally slopes at 1:400+.

2. Second, for carrying water for irrigation purposes:

    a. On hilly land, dug into ground, slopes at 1:400+.

    b. On flat land, generally flat, built above land with two banks, called the Flood-flow irrigation channel.

    c. The Irrigation Channel is an important artificial water line. Above it is rain pasture, below it is irrigated pasture.

    d. Related water control lines are steering banks, perpendicular to contour.

3. Drainage ditches are also water channels, but they are not central to Keyline.

The Keypoints of successive primary valleys will often have a rising/falling relationship.

Keypoint dams can be connected by diversion channels. If the fall of the diversion is less than the fall of the water drainage line (stream) an increasingly large area of land will be irrigable between the dams and the water drainage line. We design accordingly.

Irrigation

Hillside Irrigation: Keyline Pattern Irrigation. Flood irrigation of hilly land made possible by Keyline Cultivation Pattern. Water is stored in large dams, released through large pipes in the base of the dams, and is moved in irrigation channels dug into the ground. These channels have to have a fall of at least 1:300. Flags are positioned in the ditches, and spill water onto the land below the irrigation channel.

- Ploughing must be continued indefinitely to spread water evenly.

- Irrigation can be at rates of up eight acres per hour, with one person control.

Flat Land – Keyline Flood-flow irrigation. Water is stored in even larger dams, which tend to be shallower. Water is released through large (2') pipes in the base of the dam. Water is moved in channels which are located above the surface of the land. The channel is generally level. Gates in the channels are opened, and water spreads in a wide sheet across the land in irrigation bays. Irrigation bays are bounded by "steering banks," which run perpendicular to contour.

- Water can be applied at 20-50 acres/hour.

- Cultivation need only happen during the soil-building conversion period of three years.

Traditional Irrigation:

- Border Check Irrigation similar to flood flow, but slow.

- Contour Bay Irrigation Rice Paddies.

- Furrow Irrigation common for vegetables and orchards.

- Spray Irrigation common, expensive, lots of machinery.

- Drip tape vegetables. Not a broad-acre strategy. Lots of plastic.

Slow irrigation drowns soil aerobes. Slow irrigation is not generally sustainable.

No conventionally irrigated civilization has ever survived.

Roads

Roads on contour require less energy to travel. They do not erode easily or concentrate run-off. Roads are built in relation to water control lines.

Possible Locations of Roads

- Along boundary lines. Generally not on contour, often difficult to maintain, tend to self-destruct. Useful for fence maintenance.

- On ridge crests (watershed lines, main ridge centres). High, dry, easy to maintain. Good site for a main road.

- Located by water channels: diversion channels, irrigation channels, irrigation areas.

   a. Below diversions: dry, cross dams that cross valleys.

      Above irrigation channels in hilly country bridges are often necessary.

   b. Below Flood-flow irrigation channels.

   c. At low end of irrigation area.

   d. Along streams.

Trees

Tree locations fall into place when the first four factors have been considered. Clearing of trees and planting of trees should be considered in light of the four first landscape design considerations.

Contour Strip Forests generally follow the patterns of water harvesting/distribution channels, as well as the roads. Trees usually border roads, and are located above irrigation channels. It is good to plant trees along riparian corridors and around lakes and ponds. Pasture and crop land are separated by contoured tree lines. In the long run, trees do not interfere with productive crop land, they enhance it:

- Trees act as mineral pumps,

- Trees reduce the effects of wind,

- Trees give edge effect,

- Trees can be designed to provide browse,

- Trees provide wildlife habitat,

- Shelter.

Contoured timber belts in hill country are generally spaced so that the top of the mature trees will be level with the base of the next higher belt of trees.

Keyline soil development on pastureland prior to tree establishment will accelerate tree growth. Build soil fertility first.

Buildings

Building should be placed to optimise the potential energy flows eg:

- Not too exposed. The best view is a often a costly one from an energy consumption perspective.

- Good solar access to enable energy efficient house and building design.

- Topographic protection from prevailing wind direction.

- Build your shed higher than the house so as to use the shed water tank for gravity-fed water to the house.

- On a slope to allow good air & water drainage, gain gravity potential & out of danger from floods.

Fences

Follows all of the other infrastructure layout. Many paddocks are good. Temporary fence offers flexibility. Fences are built according to natural and artificial water lines. Rule of thumb is to build fences:

a. Along creeks, drainage lines and main ridge crests so as to create drainage line protection and to connect allow flows of wildlife from the bottom to the top of landscapes.

b. Lightweight electric internal fencing according to stock type for planned or management intensive grazing.

c. All dams and open water bodies to remove stock access.

d. Along shelterbelts, strip forests, forest plantations & revegetation forests or areas of natural significance that need protection.

Soil

Subsoil can be quickly turned into topsoil. Development & Maintenance of Soil fertility is a product of management.

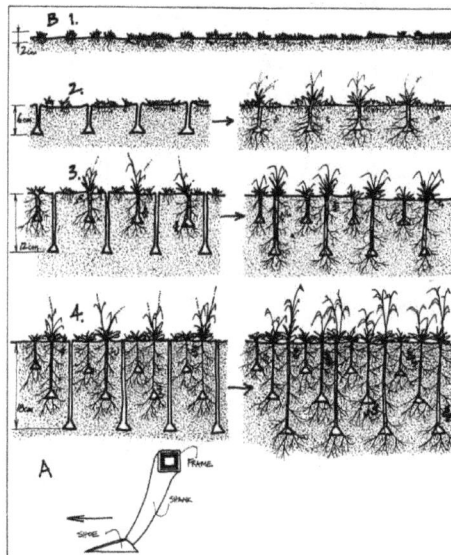

Good grazing gives the greatest return for the least energy input in increasing soil fertility. The subsoiler greatly accelerates normal topsoil formation under pasture. Conversion of subsoil to topsoil involves creating repeated biological climaxes. Soil life requires air, moisture, warmth, space & plenty of high energy, high protein food. Create these conditions, & soil life will respond, transforming some portion (often about 10%) of plant exudates and sloughed grass roots into humus. Create these conditions repeatedly, and subsoil will be "permanently" transformed into topsoil.

Urban Design

The 'Keyline Scale of Permanence' can be applied to urban design in a way that insures that water supply is clean and perpetual, transport uses minimal energy, as roads are located on or close to contours, wastewater is used to "irrigate" city forests.

- Most useful in the design of new cities.

- Dams are located with water lines at Keyline.

- Roads are laid out in relation to water control lines.

- Cities are designed from the crest of main ridges downward.

- Trees are planted/left in relation to water control lines.

- City Forests provide cleansing and valuable construction materials.

Windbreak

Windbreaks are barriers used to reduce and redirect wind. They usually consist of trees and shrubs but also may be perennial or annual crops and grasses, fences, or other materials. The reduction in wind speed behind a windbreak modifies the environmental conditions or microclimate in this sheltered zone.

As wind blows against a windbreak, air pressure builds up on the windward side (the side toward the wind), and decreases on the leeward side (the side away from the wind). Some of the approaching wind flows through the windbreak, some goes around the ends, but most of it is forced up and over the top of the wind- break. Windbreak structure — height, density, number of rows, species composition, length, orientation and continuity — determines which path the wind will take, and as a result, determines how effective the windbreak will be in reducing wind speed and altering microclimate

A well-designed farm or ranch incorporates many types of windbreaks
to protect fields, livestock, and the homesite.

Windbreak Characteristics

Effect of Height, Length and Continuity

Windbreak height (H) is the most important factor determining the distance downwind protected by a wind- break. This value varies from windbreak to windbreak and increases as the windbreak matures. In multiple row windbreaks, the average height of the tallest tree row determines the value of H. Although the height of a windbreak determines the extent of the protected areas, the length times the extent determines the total area receiving protection. For maximum efficiency, the unin- terrupted length of the windbreak should be at least 10 times its height.

The continuity of a windbreak also influences its efficiency. Gaps in a windbreak become funnels that accelerate wind flow, creating areas on the downwind side of the gap in which wind speeds often exceed open field wind speeds. Where gaps occur, the effectiveness of the windbreak is diminished. Lanes or field access should be located at the end of a windbreak. If a lane must go through a windbreak, it should be located such that the opening is at an angle to problem winds.

On the windward side of a windbreak, wind speed reductions are measurable upwind for a distance of two to five times the height of the windbreak (2H to 5H). On the leeward side, wind speed reductions occur up to 30H downwind of the barrier. For example, in a windbreak where the height of the tallest tree row is 30 feet, lower wind speeds are measurable for 60 to 150 feet on the windward side and up to 900 feet on the leeward side. The magnitude of the wind reduction at any location in the protected zone is determined by the structure of the windbreak.

Effect of Windbreak Structure

Windbreak structure is made up of two components: internal structure — the amount and arrangement of the solid elements and open spaces; and external structure — the cross-sectional shape of the windbreak.

The internal structural characteristics of a wind- break, especially the amount and arrangement of the surface area and volume of the trunk, branches, and leaves or needles, determine the magnitude of wind speed reductions. In practice, this internal structure is simply described in terms of density.

Windbreak density is the ratio of the solid portion of the barrier to the total area of the barrier. As wind flows through a windbreak, the trunk, branches and leaves (the solid portion) absorb some of the momentum of the wind and wind speed is reduced. In addition, as wind flows over the tree surfaces, it is slowed by the rough- ness of the surface and wind speed is reduced. Together, these two processes help determine the amount of wind speed reduction that occurs.

Around very dense windbreaks, air pressure builds up on the windward side and a zone of low pressure develops on the leeward side. The windward air pressure pushes air through and over the windbreak, while the leeward low pressure area behind the wind-break pulls air coming over the windbreak downward, creating turbulence and reduc-ing protection downwind. As density decreases, the amount of air passing through the wind- break increases, moderating the pressure differences between the windward and leeward sides and reducing the level of turbulence created by the dense windbreak. As a result, the extent of the downwind-protected area increases. While the extent of this protected area is larger, the wind speed reductions are not as great as those leeward of a more dense windbreak. By adjusting windbreak density, different wind flow patterns and areas of protection can be established.

The species used and their arrangement, the number of rows and the distance between rows, and the distance between trees are the main factors controlling windbreak densi-ty. Increasing the number of windbreak rows or decreasing the distance between trees increases density and provides a more solid barrier to the wind. Conifer species, such as cedar and pine, and shrubs with multiple stems tend to provide better year-round density, while taller hardwood species, such as ash, oak, or hackberry, generally are used to provide greater height.

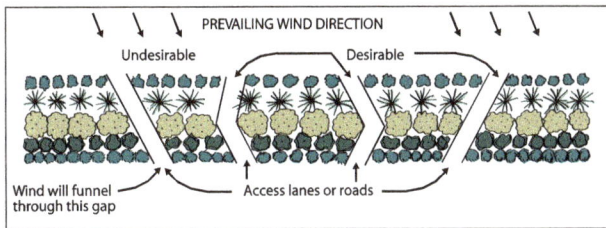

Acess lanes and roads should be at an angle to prevailing or troublesome winds.
In areas where snow is a concern snow drifts may block lane access.

The interaction of height and density determines the degree of wind speed reduction and, ultimately, the extent of the protected area. For a windbreak with a given height, the length of the protected area downwind usually increases as density increases from 20 to 60 percent. At densities below 20 percent, the windbreak provides little, if any, useful wind reduction. As densities increase above 60 percent, leeward turbulence be-gins to increase, the length of the protected area downwind begins to shrink and wind-break efficiency is decreased.

The external structure or cross-sectional shape of a windbreak is determined by the width, height and arrangement of the individual tree and/or shrub rows within the windbreak. The cross-sectional shape of windbreaks with similar internal structures has minimal influence on wind speed reductions within 10H of the barrier. Beyond 10H, windbreaks with a vertical windward side tend to provide slightly more protection than those with a slanted windward side, because more wind passes through the barrier reducing turbulence and extending the protected area farther to the lee.

Windbreaks with a streamlined shape in cross-section, similar to a gabled roof, have been advocated in the past. This usually is achieved by planting central rows with tall trees and flanking both sides with shorter trees or shrubs. In most cases, this design is less efficient, requiring more land but not necessarily providing increased wind protection. However, these wider wind- breaks provide valuable wildlife habitat benefits and are an appropriate design when wildlife habitat is an important objective of the landowner.

Windbreak Design

In designing a windbreak, density should be adjusted to meet the landowner's objectives. In general, wind- breaks with higher densities (multiple rows) are used to protect wildlife, farmsteads, or homesites, while wind- breaks with lower densities (one or two rows) are used to protect crop fields.

A windbreak density of 40 to 60 percent provides the greatest downwind area of protection and provides excellent soil erosion control. To get uniform distribution of snow across a field, densities of 25 to 35 percent are most effective, but may not provide sufficient density to control soil erosion. Windbreaks designed to catch and store snow in a confined area usually have three to five rows of conifers or shrubs and densities in the range of 60 to 80 percent. Farmsteads and livestock areas needing protection from winter winds require multiple row windbreaks with high densities. Typically, these windbreaks have two or three rows of conifers, one or two rows of tall hardwoods, and one or more rows of shrubs. In these cases, wind speed reductions are greater but the extent of protected area is smaller.

Open Wind Speed 20 mph Deciduous 25-35% Density					
H distance from windbreak	5H	10H	15H	20H	30H
miles per hour	0	13	16	17	20
% of open wind speed	50%	65%	80%	85%	100%
Open Wind Speed 20 mph Conifer 40-60% Density					
H distance from windbreak	5H	10H	15H	20H	30H
miles per hour	6	10	12	15	19
% of open wind speed	30%	50%	60%	75%	95%
Open Wind Speed 20 mph Multi Row 60-80% Density					
H distance from windbreak	5H	10H	15H	20H	30H
miles per hour	5	7	13	17	19
% of open wind speed	25%	35%	65%	85%	95%
Open Wind Speed 20 mph Solid Fence 100% Density					
H distance from windbreak	5H	10H	15H	20H	30H
miles per hour	5	14	18	19	20
% of open wind speed	25%	70%	90%	95%	100%

Wind speed reductions to the lee of windbreaks with different densities. A)
density of 25-35%, B) density of 40-60%, C) density of 60-80%, D) density of 100%.

Windbreaks are most effective when oriented at right angles to prevailing winds. The purpose and design of each windbreak is unique; thus, the orientation of individual windbreaks depends on the design objectives.

Farmsteads and feedlots usually need protection from cold winds and blowing snow or dust. Orienting these windbreaks perpendicular to the troublesome winter wind direction provides the most useful protection. This usually is accomplished by planting windbreaks on the north and west sides of the farmstead or feedlot.

Successful field windbreaks should be designed to fit within the farming operation. Consideration should be given to reducing wind erosion, providing crop protection, increasing irrigation efficiency and improving wildlife habitat.

Field crops usually need protection from hot, dry summer winds; abrasive, wind-blown soil particles; or both. The orientation of these windbreaks should be perpendicular to prevailing summer winds, usually south or west. Windbreaks designed to protect fall- seeded small grains like winter wheat may need protection from both summer and winter winds. To control soil erosion, windbreaks should be planted to block prevailing winds during the times of greatest soil exposure — usually winter and early spring. To recharge soil moisture with drifting snow, windbreaks should be placed perpendicular to prevailing winter winds.

Although wind may blow predominantly from one direction during one season, it rarely blows exclusively from that direction. As a result, protection is not equal for all areas on the leeward side of a windbreak. As the wind changes direction and is no longer blowing direct- ly against the windbreak, the protected area decreases. The use of multiple-leg windbreaks provides a more consistent and larger protected area than a single windbreak. Again, individual placement depends on the site, wind directions, and design objectives.

Microclimate Modifications

The reduction in wind speed adjacent to a windbreak reduces upward transport of heat and moisture from the soil surface. As a result, temperature and humidity levels in the sheltered zones usually increase and evaporation and plant water loss decrease. These changes contribute to conservation of soil moisture, improvement of crop water use efficiency and an increase in crop yields in the protected zone.

Actual temperature modifications for a given wind- break depend on windbreak height, density, orientation, and time of day. Daily air temperatures within 10H leeward of a windbreak are generally several degrees higher than temperatures in the open. Beyond 10H, air temperatures near the ground tend to be slightly cooler during the day. On most nights, temperatures near the ground in sheltered areas are slightly warmer than in the open due to the reduction in wind speed and in the upward transfer of heat from the surface. In contrast, on nights when wind speeds are very low, the reduction in wind speeds in shelter may lead to greater levels of radiation cooling and sheltered areas may be several degrees cooler than open areas. In early spring and late fall, these conditions may lead to frost in sheltered areas.

Early in the growing season, soil temperatures in sheltered areas usually are several degrees warmer than in unsheltered areas. Taking advantage of these warmer temperatures may allow earlier planting and more rapid germination in areas with short growing seasons. In the area immediately adjacent to an east- west windbreak, soil temperatures tend to be higher on the south side due to heat reflected off the windbreak. On the north side, soil temperatures, especially in the early spring, are lower due to shading by the windbreak. These cooler temperatures reduce the rate of snow melt, and, in more northern areas, may cause problems with field access in early spring.

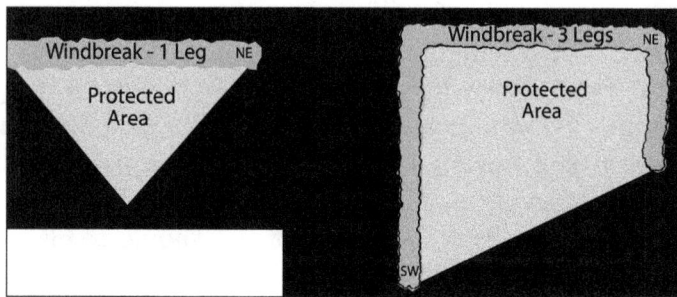

In areas with winds from many directions, multiple-leg windbreaks or windbreak systems provide greater protection to the field or farmstead than single-leg windbreaks.

Relative humidity in sheltered areas is two to four percent higher than in open areas. Higher humidity decreases the rate of plant water use, so production is more efficient than in unsheltered areas. However, if the windbreak is too dense and humidity levels get too high, diseases may become a problem in some crops.

Moderation of windchill is most important in farm- stead and livestock windbreak situations where humans and other animals readily notice the effects of cold winter winds. Livestock use less feed and suffer less weather related stress when protected from winter winds. Similarly, good winter protection for outdoor work areas makes winter chores less stressful and reduces the risk of injury due to extreme cold.

References

- Hydroculture: appropedia.org, Retrieved 19 April, 2019

- What-is-hydroculture: hydroculture.co.uk, Retrieved 9 January, 2019

- Hydroculture-growing-plants-without-soil: ambius.com, Retrieved 2 July, 2019

- Aeroponics: maximumyield.com, Retrieved 22 January, 2019

- Crossley, Phil L. (2004). "Sub-irrigation in wetland agriculture" (PDF). Agriculture and Human Values. 21 (2/3): 191–205. doi:10.1023/B:AHUM.0000029395.84972.5e. Archived (PDF) from the original on December 6, 2013. Retrieved April 24, 2013

- Aeroponics-benefits-and-disadvantages, aeroponics: gardeningsite.com, Retrieved 11 May, 2019

- Hydroponics: maximumyield.com, Retrieved 4 August, 2019

- What-is-hydroponics: aquagardening.com.au, Retrieved 14 February, 2019

- Advantages-disadvantages-of-hydroponics: greenandvibrant.com, Retrieved 12 April, 2019

- What-is-regenerative-agriculture: regenerationinternational.org, Retrieved 22 July, 2019

- Integrated-pest-management: maximumyield.com, Retrieved 19 May, 2019

- Integrated-pest-management: extension.org, Retrieved 29 March, 2019

- Biological-pest-control: newworldencyclopedia.org, Retrieved 15 February, 2019

- Companion-planting-in-gardening, cooperative-extension: agriculture.vsu.edu, Retrieved 25 June, 2019

- What-is-permaculture-farming: greentumble.com, Retrieved 3 January, 2019

- What-is-permaculture: tenthacrefarm.com, Retrieved 13 May, 2019

- What-is-permaculture: neverendingfood.org, Retrieved 19 February, 2019

- Permaculture-gardening-techniques: greenglobaltravel.com, Retrieved 29 August, 2019

- Key-line-soil-biology: aquinta.org, Retrieved 30 March, 2019

- Howwindbreakswork: nfs.unl.edu, Retrieved 12 June, 2019

Chapter 4

Sustainable Agricultural Practices

There are a number of practices which can make agriculture more sustainable by reducing long-term damage to the soil. A few of them are crop rotation, planting cover crops, drip irrigation, green manuring and multiple cropping. This chapter discusses in detail these agricultural practices.

Crop Rotation

Crop rotation refers to the sequence of crops grown in a specific field, including cash crops, cover crops and green manures. Rotations are the changing of crops over both space and time. Well-planned rotation schedules benefit soil fertility, aid in pest management, spread labor needs over time and reduce risks caused by market conditions. Factors such as crop family, plant rooting depths and crop fertility needs should be considered when developing a crop rotation schedule.

Reasons to rotate:

- Enhances soil quality,
- Increases soil fertility,
- Aids in pest management.

Soil Quality

Crop rotation practices such as manuring, composting, cover cropping, green manuring and short pasturing cycles improve soil quality by maintaining or increasing soil organic matter content. Organic matter serves as the primary food source for soil microorganisms. These organisms provide many benefits, including holding the soil particles together, releasing minerals for plant uptake, enhancing the downward movement of water and air, and providing pathways for root growth. Rotations including crops with a variety of rooting depths make use of water and nutrients throughout the soil, aid in loosening compacted soil and increase topsoil over time.

Fertility

Well-planned rotation schedules take into account the preceding year's crops, ensuring that nutrients are available for crops grown the following season.

It is important to consider the nutrient needs of each crop to ensure they will be met. The addition of leguminous crops in a rotation can provide nitrogen for following crops. Including crops with a variety of rooting depths allows crops to retrieve water and nutrients not accessed by those grown in previous rotations. Some plants are also effective at making nutrients more available by using less soluble forms, making them accessible for later crops.

Pest Management

Pests are most easily kept in balance when different crops are grown over a number of years. Rotate susceptible crops at intervals to inhibit the buildup of their specific pest organisms. Rotation length should be based on the amount of time soil-borne pathogens remain viable in the field. A four-year rotation using crops not susceptible to the same pathogens will generally minimize problems from soil-borne pathogens, with some exceptions. Two years is considered enough time to reduce the incidence of foliar diseases.

Cropping sequence should be determined based on susceptibility to insect pests. Succeeding crops should have different growth habits and be host to a different set of pests. The primary goal in managing insects through crop rotation is to interfere with the needs of the pest throughout its life cycle. It is therefore important to be familiar with insect life cycles, feeding habits and crop preferences.

The best method of weed control is optimizing crop growth to reduce niches for weeds to develop. Crop rotation helps suppress weeds by using crops that out- compete weeds for water, nutrients or sunlight. Some crops, such as rye or sorghum, release chemicals while growing or decomposing that prevent the seed germination and growth of other nearby plants; this is called allelopathy. The use of cover crops during non-production periods can decrease weed pressure by allelopathy or competition and, when killed and left as a mulch, cover crops can suppress weeds by shading the soil surface.

Rotation strategies:

- Rotate by plant family.

- Rotate by plant part harvested.

- Rotate by plant compatibility.

- Rotate by nutrient requirements.

- Rotate by rooting depth and type.

- Include legumes and cover crops.

Rotation Method

Planning a Rotation

For ease of planning, it is good to design rotational sections of the same size. These sections can then be further subdivided based on production size and land required by each crop, or to incorporate shorter rotational cropping plans. Crops should be divided by family, so the same or closely related crops are not grown in direct succession. It may also prove beneficial to subdivide crops by cultural and management requirements, architectural structure, growth pattern, harvest date, etc. In a short-rotation system, changes should be introduced whenever possible; this may include changes in crop variety or the addition of cool-season cover crops or green manures.

Legumes

Legumes are an important addition to a crop rotation plan because they fix atmospheric nitrogen, which can be used as a replacement or supplement for inorganic nitrogen fertilizer. The total N contribution varies among species, but 50-200 lbs N/acre can be expected from a good legume cover or cash crop stand. Unlike highly soluble nitrogen fertilizers with a significant potential to leach, N supplied by legume crops can be held in soil for extended periods. Approximately 40-75 percent of the N contained in the crop may be available for subsequent plants.

Table: Rotation lengths to reduce soil-borne pathogens.

Vegetable	Disease	Yrs w/o Susceptible Crop
Asparagus	Fusarium rot	8
Cabbage	Clubroot	7
Cabbage	Blackleg	3-4
Cabbage	Black rot	2-3
Muskmelon	Fusarium wilt	5

Parsnip	Root rot	2
Pease	Root canker	3-4
Pease	Fusarium wilt	5
Pumpkin	Black rot	2
Radish	Clubroot	7

Cover Crops and Green Manures

Cover crops are also an important component of a crop rotation plan and should be utilized when fields are not being used for production. Cover crops and green manures are those crops grown specifically for the benefits they provide. They may be incorporated into the soil or left as a residue on the soil surface.

Their benefits include increased organic matter, improved soil structure, enhanced drought tolerance, increased nutrient availability for plants, protection against soil erosion, weed suppression, penetration of compacted subsoils and nutrient cycling.

Crop Families

Crops within the same family are generally susceptible to the same insect pests and diseases. A four-year rotation using crops not susceptible to the same pathogens will generally minimize problems from soil-borne pathogens, with some exceptions. Two years is considered enough time to reduce the incidence of foliar diseases and insect pests. When planning a rotation, it is often helpful to map out where the crop families listed will be located and how much of each will be planted:

- Poaceae: Corn

- Alliaceae: Onion, garlic, shallot, leeks

- Chenopodiaceae: Beet, chard, spinach

- Cucurbitaceae: Winter and summer squash, cucumber, melon, pumpkin

- Brassicaceae: Rutabaga, kale, broccoli, cauliflower, cabbage, Brussels sprouts, radish, mustard, turnip

- Fabaceae: Pea, bean

- Apiaceae: Carrot, parsley, celery, parsnip Solanaceae: Potato, tomato, pepper, eggplant Asteraceae: Lettuce

- Convolvulaceae: Sweet potato

- Malvaceae: Okra.

Compatability

It is important to consider crop compatibility when planning a rotation. Some crops may have beneficial interactions and enhance yield, while others may have detrimental effects to subsequent crops. For example, many crops following the cabbage family may have lower yields. Sweet corn is a good selection to follow the cabbage family because it shows no yield decline. Potatoes are a good crop to follow sweet corn because research has shown sweet corn to be one of the preceding crops that most benefit the yield of potatoes.

Factor Affecting Crop Rotation

- Climate:

  Climate is the one of most important factor which is effect the crop rotation either by wind, rain or other factors.

- Type and nature of soil:

  Type and nature of soil is also important factor which effects the crop rotation some soil are fertile and some are low in fertility

- Availability of inputs:

  Availability of inputs at the place is also effects the crop rotation like fertilizer, pesticide etc

- Availability of labor:

  Availability of labor is effect the crop rotation. The labor is required at the critical stages of crop if the labor is not available at that time the crop may cause loss

- Situation of farm:

  The farm location is also very important factor which is effect the crop rotation.

- Size of Farm:

  The size of farm is effects the crop rotation. Small land holding is major problem in Pakistan that's why crop rotation is effect by the farm size.

Planting Cover Crops

Cover crops are crops grown to improve the farming system. Cover crops are typically planted between rotations of income-producing crops, but they can also be planted at the same time. Cover crops fulfill a wide variety of management objectives and serve as integral components of organic farming systems. There are many species of cover crops to choose from.

A polyculture of crimson clover, cereal rye and hairy vetch used as a green manure cover crop for sweet corn. Planting polycultures increases plant and insect diversity, increases the number of ecosystem services, and decreases risk of crop failure.

Integrating cover crops can have significant ecological impacts on the farming system. Cover crops can improve soil physical, chemical, and biological properties; supply nitrogen; reduce leaching of nutrients and pesticides; reduce erosion; mitigate damage from plant pests and reduce their population densities; as well attract beneficial insects. Cover crops can also generate additional income when grown for seed or feed, or as an energy crop. While it is difficult to achieve all of the listed benefits with one crop, producers should select cover crops that offer multiple benefits at once. Producers should also consider potential drawbacks before deciding to include a cover crop. In some instances, the cover crop can require additional labor and expense, delay crop planting, or serve as an alternate host for crop insects or diseases.

A monoculture stand of pearl millet 'Tifleaf3' one week before termination. This stand was seeded at a rate of 25 pounds per acre and produced 3,750 pounds per acre of biomass on a dry weight basis.

A cover crop should:

- Satisfy the producer's primary reason(s) for cover cropping;

- Be easy to establish and maintain with available equipment;

- Be well-suited to the local climate and the farm environment;

- Not compete with income-generating crops grown simultaneously or subsequently; and

- Have the ability to withstand stresses likely to occur such as drought, frost, heat, etc.

Prioritizing objectives for cover crops necessitates an understanding of when and under what conditions benefits can occur. Some benefits occur during cover crop growth, while other benefits occur after cover crop termination. Generally, benefits are only fully realized with a robust stand of cover crop. A wide variety of cover crop species and management options are available to fit a farm operation. Producers have many options in species selection and management, and the selection and management of species will be dictated by producer needs and production constraints.

Cover Crop Benefits

Cover crops have a surprisingly wide array of benefits and no serious drawbacks. A cover crop can improve the health of your soil, resulting in a significantly larger, healthier cash crop for the next growing season. Cover crops:

- Improve biodiversity by increasing the variety of species in a given area. For example, if there are more, varied insects that feed on the vegetation, it can bring more birds and so on.

- Reduce the amount of water that drains off a field, protecting waterways and downstream ecosystems from erosion. Because each root of the cover crop creates pores in the soil, cover crops help allow water to filter deep into the ground. As a result, a cover crop can help conserve water and prevent soil erosion.

- Help break disease cycles by reducing the amount of bacterial and fungal diseases in the soil. If you have a soil that is infested, you can plant a cover crop in that area as a means to eradicate the disease.

- Provide nutrients to the soil, much like manure does. They are also called "living mulches" because they can prevent soil erosion. Mulch is a layer of organic material, such as crop residue, that is left on the surface of the soil to prevent water runoff and protect the soil from the damaging effects of heavy rainfall.

Organic Gardening with Cover Crops

Cover crops are an important part of sustainable agriculture. These crops add fertility to the soil without chemical fertilizers via biological nitrogen fixation. A cover crop can offer a natural way to reduce soil compaction, manage soil moisture, reduce overall energy use, and provide additional forage for livestock.

Small farmers choose to grow specific cover crops based on their needs and goals and the overall requirements of the land they are working. Cover crops grown in summer are often used to fill in space during crop rotations, help amend the soil, or suppress weeds. Winter cover crops help hold soil in place over the winter and provide ground cover. These crops can also fix nitrogen levels in the soil.

Planting after a Cover Crop

Once a cover crop is fully grown, or the farmer wants to plant in an area that has a cover crop, the conventional technique is to mow down the cover crop and allow it to dry. After it is dry, the remaining organic matter is usually tilled into the soil. Alternatively, some progressive farmers in drought-prone areas favor a no-till method, in which the residue from the cover crop is left on the soil as a mulch layer.

Types of Cover Crops

Examples of plants that have proven to be effective cover crops include:

- Rye: Also known as winter rye or cereal rye, this cover crop is often used to loosen compact soil and suppress weeds.

- Buckwheat: Fast-growing buckwheat helps prevent erosion and suppress weeds.

- Clover: Clover is great for fixing nitrogen in the soil and adding fertility.

- Sorghum: This hybrid cover crop grows quickly, adds biomass, and suppresses weeds.

- Hairy vetch: Vetch adds nitrogen and is a good overwinter crop for northern climates.

Implementation of Cover Crops

Cover crops are integrated into organic farming systems in many ways. They are used in rotation in vegetable and dairy pasture systems, as living mulches, as green manures, and as a mulch on the soil surface. Cover crops provide many benefits, but there is no single cover crop that will fullfill all your requirements in every situation.

1. Identify the main objectives:

Producers must first prioritize their reasons for planting a cover crop. There are a number of benefits cover crops can provide. These include:

- Provide nitrogen;

- Add organic matter;

- Improve soil structure;

- Reduce soil erosion;

- Provide weed control;

- Manage nutrients;

- Provide food for pollinators; and

- Furnish moisture-conserving mulch.

You might also want the cover crops to provide habitat for beneficial organisms, better traction during harvest, faster drainage, or another benefit. A few species known to provide these additional benefits should be tested in the farming system, prior to planting a large area.

Figure: A biculture of black oats and crimson clover are drilled in alternating rows as a green manure cover crop for sweet corn.

2. Identify the best place and time:

Sometimes it's obvious where and when to use a cover crop. For example, you might want plant a legume to add some nitrogen before a corn crop, or plant a perennial ground cover in a vineyard or orchard to reduce erosion or improve weed control. If secondary goals such as soil building are an objective, then it may become more difficult to decide where and when to schedule cover crops. To plan how and where to use cover crops, try the following exercise:

Look at your rotation. Make a timeline of 18 to 36 monthly increments across a piece of paper. For each field, pencil in current or probable rotations, showing when you typically seed crops and when you harvest them. If possible, add other key information, such as rainfall, frost-free periods, and times of heavy labor or equipment demand. Look for open periods in each field that correspond to good conditions for cover crop establishment, under-utilized spaces on your farm, as well as opportunities in your seasonal work schedule. Also, consider ways to extend or overlap cropping windows.

3. Select the best cover crop:

A goal, a time, and a place; now specify the traits a cover crop would need to work well.

4. Settle for the best available cover:

It's likely the "wonder crop" you want doesn't exist. One or more species could come close. Top regional cover crop species can provide a starting point. Keep in mind that you can mix two or more species, or try several options in small areas.

5. Build a rotation around cover crops:

It's hard to decide in advance every field's crops, planting dates, fieldwork, or management needs. One alternative is to find out which cover crops provide the best results on your farm, then build a rotation around those covers, especially when trying to tackle some tough soil improvement or weed control issues. With this "reverse" strategy, you plan cover crops according to their optimum field timing, and then determine the best windows for cash crops. A cover crop's strengths help you decide which cash crops would benefit the most.

Reduced Tillage

Reduced tillage or conservation tillage is a practice of minimising soil disturbance and allowing crop residue or stubble to remain on the ground instead of being thrown away or incorporated into the soil. Reduced tillage practices may progress from reducing the number of tillage passes to stopping tillage completely (zero tillage).

It is becoming popular because of the direct economic benefits it provides farmers. With less tilling, farmers save on machinery use, fuel, labour and their own time. No tillage is also an important part of natural farming popularised by Masanobu Fukuoka, It is possible on any chemical-free farm with a balanced agro-ecosystem and sensible cropping practices.

Reducing tillage is important from the viewpoint of environmental-farming for a number of reasons. The cover of crop residue helps prevent soil erosion by water and air, thus conserving valuable top soil. Soil structure improves because heavy machinery (which causes soil compaction) is not used and soil tilth is not tampered with artificially. With earthworms not being routinely disturbed by deep tillage, their numbers increase bringing with them the accompanying benefits of better soil aeration and improved soil fertility. Microbial activity in soil also increases for the same reason. Another important environmental effect of reduced tillage is the reduction in use of fossil fuels on the farm.

The excessive tillage that occurs on most vegetable farms (plowing, harrowing, culti-packing, bed formation, cultivating) has many unintended consequences for soils and the environment. Some of the problems associated with excessive tillage include; loss of organic matter and beneficial soil organisms, increased soil erosion and pesticide runoff, reduced soil fertility, loss of soil structure and porosity, compaction, surface crusting, formation of plow pans, reduced root growth, poor drainage, and reduced water-holding capacity. Results from a recent survey of 55 vegetable farms in Connecticut found that almost 90% of conventionally tilled vegetable farms had plow pans, compared with 33% for reduced-till operations, while the latter group had almost twice as much organic matter in their soils.

Tillage is also expensive and consumes a lot of energy. Reduced-tillage systems can often reduce fuel usage and reduce field preparation time by over 66% when compared with conventional tillage systems. These systems can provide equal or better yields than conventional tillage and may provide many other benefits as well.

Reducing the amount of tillage that takes place can help reverse the problems associated with excess tillage and begin to restore the health of a soil. A simple way to reduce tillage on your farm includes swapping from moldboard plows, disk-harrows and roto-tillers to using less impactful implements like chisel plows, subsoilers, s-tine cultivators and spaders. You may also work towards implementing minimum tillage systems such as strip-till, zone-till, ridge-till, no-till or permanent-bed systems. Most reduced-till systems are used in conjunction with cover crops or organic mulches to protect the soil surface at all times, help increase organic matter over time, or to help control weeds. Other examples of ways to reduce tillage include:

1. Using chisel plow shanks, subsoilers or zone-tillers to loosen soil before preparing raised-beds instead of a plow and harrow;

2. Planting summer cover crops, such as buckwheat, after an early cash crop, as a substitute for repeated harrowing to control weeds;

3. Mowing crop residues instead of disking;

4. Planting tillage radishes or other deep-rooted cover crops to help prevent plow pans from reforming;

5. Using a no-till drill to plant cover crops, instead of a harrow to assure good seed-to-soil contact for emergence.

Deep zone-tillage, also known as vertical-tillage, is one of the more promising and versatile methods of reduced tillage for vegetables in our climate and can help vegetable farmers reverse the ill effects of years of excessive tillage on their soils. Deep zone-tillage is similar to no-till in that it relies on the residue of a cover crop to protect the soil surface and help improve soil health over time. However, no-till relies on a heavy blanket of plant residue in the planting row to protect the soil, and inadvertently delays crop growth by keeping soils in the root zone cool in Northern climates. Deep zone tillage addresses this issue by incorporating a 5 to 12"-wide tilled strip to simultaneously break up plow pans, prepare seedbeds and warm the soil. Planting and fertilizing can often be done in the same pass, further reducing fuel, machine hours, labor costs, fertilizer rates, and soil compaction. Soil drainage can be improved immediately and continues to improve each year. The same herbicides, or some of the same cultural practices, are used to control annual weeds.

Implements used for deep zone-tillage usually consist of a lead coulter to cut through the killed-cover crop residue, followed by a deep shank or subsoiler to break up the plow-pan, and finally a pair of fluted coulters and a rolling basket to prepare a narrow seedbed and help break up soil clods. The deep shanks are mounted onto a hinged frame, which allows the shanks to rise out of the ground when they encounter large rocks or ledge, while spring resets push the shanks back down into position after passing over the obstacle. Crop roots grow deep through the slit made by the shank rather than just spreading out in the top few inches of soil above the plow pan. Additional coulters or (finger-like) residue managers are mounted on the planter in front of the planting shoes to remove excess cover crop residue and stones to provide finished seedbeds.

The soil surface between the crop rows retains the heavy surface residue from the dead cover crop. The 5 to 12"-wide tilled strip warms faster than residue-covered soils and, if installed across a slope, does not allow water to build up enough speed to erode a slope. Roots and surface residue from the cover crop in the untilled area between crop rows do not break down as fast as when the soil is tilled/aerated, so organic matter tends to increase over time. With the return of organic matter, comes the return of beneficial soil organisms, better soil structure, better water infiltration and holding capacity, and a healthier, more productive soil.

There are challenges to successful zone tillage management. Killing cover crops and weed control can be problematic, especially with organic systems that do not allow herbicide use. Plant establishment can also be negatively affected by the presence of cover crop residues. Growers will need to be innovative to overcome these challenges. Organic growers may try planting perennial rye or turf grass in the fall, and using a modified rototiller with the outside tines removed, to prepare narrow strips or seedbeds at the desired row spacing in the spring. A subsoiler could be used to rip through the plow pan under the prepared strip to improve drainage. The living grass mulch between the crop rows can be controlled by mowing, while weeds within the row could be controlled by mulching, flaming or hoeing, or by planting competitive crops, such as summer squash. At the end of the season, simply seed the strip back to turf. The next season, move the strips mid-way between the previously prepared rows and switch crops to complete your crop rotation.

There are other options to avoid using herbicides. Before early-planted spring crops, use fall-planted oats or a blend of cover crops that winter-kill, such as oats and tillage radish, before zone tilling and planting. Cultivation can then be used for weed control over the relatively broken down cover crop residue. For summer vegetable plantings, use winter rye, but wait until it sheds pollen in June to crush it with a roller crimper, or cut and bail or windrow it to help suppress weeds between rows. Organic farmers who have worked most of the weed seed bank out of the top few inches of their soil through a combination of winter/summer cover crops, mulches, summer fallow periods and timely cultivations, may find it easier to adapt to zone tillage than those fighting high levels of weed seeds in their soils. Note that specialized cultivation equipment will be needed to manage in-row weeds. The heavy residue or living mulch between rows will make mid-season cultivation of those areas difficult. Be sure to try this practice on small areas with low weed pressure. Small farms with equipment of insufficient size to pull a zone tiller, might try lighter weight equipment to break through a plow pan and produce a seedbed, such as a Yeomans Plow, which can be pulled with a 16 to 18 horsepower tractor.

Strip-tillage, sometimes referred to as shallow zone-tillage, is similar to deep zone tillage without the subsoiling shank to break up the plow pan. The implement has two or three closely spaced coulters and a rolling basket to prepare and smooth a narrow seedbed through the surface residue. Because the implement lacks a deep shank, this system does not have the ability to improve drainage immediately, and it may take several years for the soil health attributes and drainage to improve. However, on farms without a plow pan this system can provide most of the benefits of deep zone tillage and uses less fuel.

Strips for cash crops can vary in width. To make wider strips in winter rye and vetch for late-planted vegetables, use a spader to prepare planting strips early in the spring when the rye just begins to grow, leaving equally wide strips of the cover crop to mature. A cultipacker or some other finish tool may be needed to smooth the seedbed for small seeded crops. Plant or transplant the cash crop in the prepared beds while the cover crop continues to grow between the beds. To kill the cover crop, cut it when the rye is shedding pollen or when the vetch begins to flower, and spread the straw residue over the prepared bed as a mulch to help suppress weeds around the cash crop. It is best to cultivate the beds once before cutting and spreading the residue from the adjacent cover crops. Supplement the rye/vetch mulch with straw from a nearby field of rye.

For early season vegetables, use a two-year system with spring-planted oats and field peas. During the first summer, after the cover crop forms seeds, mow it to get a thicker stand late in the season. After the cover crop winter-kills, use a spader to make planting beds for vegetables and use a straw mulch between beds or cultivate. A similar process can be done with medium red clover sown between or under a cash crop the first year. When the cash crop is harvested and mowed off in late summer or fall, the clover will fill in to make a solid stand by spring. A spader can then be used to make seedbeds for the new cash crop in the clover stand.

No-Till planters have double-disk openers and closing wheels to create and close the seed furrow in unworked soil, through a thick cover crop residue. These planters rely on down-pressure springs and/or extra weight to assure that the seed furrow can be created, especially in a dry or compacted soil. If the accumulated crop residue is too thick or unevenly distributed the planters may also have residue managers to move some of the debris before planting. No-till planting can be used for late-planted vegetables in New England, after the soil has warmed under the cover crop residue. It works well for pumpkins and winter squash or summer plantings of sweet corn or other vegetables. When transitioning from conventional to no-till, yields have been known to decline slightly for a few years before recovering as the soil characteristics improve.

Adoption in the United States

No-till farming is widely used in the United States and the number of acres managed in this way continues to grow. This growth is supported by a decrease in costs related to tillage; no-till management results in fewer passes with equipment for approximately equal harvests, and the crop residue prevents evaporation of rainfall and increases water infiltration into the soil.

Issues

Profit, Economics and Yield

Studies have found that no-till farming can be more profitable if performed correctly.

It reduces labour, fuel, irrigation and machinery costs. No-till can increase yield because of higher water infiltration and storage capacity, and less erosion. Another benefit of no-till is that because of the higher water content, instead of leaving a field fallow it can make economic sense to plant another crop instead.

As sustainable agriculture becomes more popular, monetary grants and awards are becoming readily available to farmers who practice conservation tillage. Some large energy corporations which are among the greatest generators of fossil-fuel-related pollution may purchase carbon credits, which can encourage farmers to engage in conservation tillage. Under such schemes, the farmers' land is legally redefined as a carbon sink for the power generators' emissions. This helps the farmer in several ways, and it helps the energy companies meet regulatory demands for reduction of pollution, specifically carbon emissions.

No-till farming can increase organic (carbon based) matter in the soil, which is a form of carbon sequestration. However, there is debate over whether this increased sequestration detected in scientific studies of no-till agriculture is actually occurring, or is due to flawed testing methods or other factors. Regardless of this debate, a case can still be made for no-till, in the form of reduction in fossil fuel use, less erosion and better soil quality.

Environmental Issues

Greenhouse Gases

No-till farming has carbon sequestration potential through storage of soil organic matter in the soil of crop fields. Tilling inverts soil layers, mixes in air, and greatly increases microbial activity. Organic matter breaks down much faster, releasing its carbon into the atmosphere. Also, farm tractors emit carbon dioxide.

Cropland soils are ideal as a carbon sink, as they have been depleted of carbon in most areas. Tillage and conventional farming have released an estimated 78 billion metric tonnes of carbon, by removing crop residues such as left over corn stalks and adding chemical fertilizers. Without tillage, residues decompose where they lie, and growing of winter cover crops can slow and reverse carbon loss.

However, there is evidence that no-till systems still lose carbon over time. A 2014 study led by Ken Olson of University of Illinois concluded that this differing result occurs in part because tested soil samples need to include the full depth of rooting, 1–2 meters. He said, "That no-till subsurface layer is often losing more soil organic carbon stock over time than is gained in the surface layer". Also, there has not been a uniform definition of soil organic carbon sequestration among researchers. The study concludes, "Additional investments in soil organic carbon (SOC) research is needed to better understand the agricultural management practices that are most likely to sequester SOC or at least retain more net SOC stocks."

Besides reducing carbon emissions, no-till farming reduces nitrous oxide (N_2O) emissions by 40-70%, depending on rotation. Nitrous oxide is a potent greenhouse gas, 300 times stronger than CO_2, and stays in the atmosphere for 120 years. Fertilizing farmlands with (excessive) nitrogen increases the release of nitrous oxide.

Soil Water

No-till farming improves soil quality (soil function), carbon, organic matter, aggregates, protecting from erosion, evaporation of water, and structural breakdown. Reducing of tillage reduces compaction of soil. This can help reduce soil erosion almost to soil production rates.

Recently, researchers at the Agricultural Research Service of the United States Department of Agriculture found that no-till farming makes soil much more stable than plowed soil. Their conclusions draw from over 19 years of collaborated tillage studies. No-till stores more carbon in the soil and carbon in the form of organic matter is a key factor in holding soil particles together. The first inch of no-till soil is two to seven times less vulnerable than that of plowed soil. The practice of no-till farming is especially beneficial to Great Plains farmers because of its avoidance of erosion.

Crop residues left intact help both natural precipitation and irrigation water to infiltrate the soil. Residue also limits evaporation, conserving water for plant growth. Evaporation from tilling reduces the amount of water by around 1/3 to 3/4 inches (0.85 to 1.9 cm) per pass. By reducing soil compaction and no tillage-pan, the soil absorbs more water, and roots grow deeper, reaching more water.

Biota and Wildlife

No-till farming leaves soil intact and crop residue on the field. Soil layers and soil biota, remain in their natural state. No-tilled fields often have more beneficial insects and annelids, more organic material and microbial content, and variety of wildlife. This is the result of improved cover, reduced traffic and the reduced chance of destroying ground nesting birds and animals. No ploughing also means less airborne dust.

Albedo

Tillage lowers the albedo of croplands. The potential for global cooling as a result of decreased Albedo in no till croplands is similar in magnitude to the biogeochemical (carbon sequestration) potential.

Historical Artefacts

Tilling regularly damages ancient structures under the soil such as long barrows. In the UK, half of the long barrows in Gloucestershire and almost all the burial mounds in Essex have been damaged. According to English Heritage modern tillage techniques have done as much damage in the last six decades as traditional tilling did in the previous six centuries. By using no-till methods these structures can be preserved and can be properly investigated instead.

Cost

Equipment

No-till farming requires specialized seeding equipment such as seed drills, to plant seeds into undisturbed crop residues and soil. The cost can be offset by selling plows and tractors, but farmers often keep their old equipment while trying out no-till farming. This would result in more money being invested into equipment in the short term (until old equipment is sold off).

Drainage

If a soil has poor drainage, it may need drainage tiles or other devices to remove excess water under no-till. Water infiltration improves after 5–8 years of no-till farming, so farmers may want to wait before investing in such an expensive system.

Gullies can be a problem in the long-term. While much less soil is displaced by no-till farming, any drainage gulleys that do form deepen each year since they are not smoothed out by plowing. This may necessitate either sod drainways, waterways, permanent drainways, cover crops, etc. Gully formation can be avoided entirely with proper water management practices, including the creation of swales on contour.

Increased Chemical use

One of the purposes of tilling is to remove weeds. No-till farming does change weed composition drastically. Faster growing weeds may no longer be a problem in the face of increased competition, but shrubs and trees may begin to grow eventually.

Some farmers attack this problem with a "burn-down" herbicide such as glyphosate in lieu of tillage for seedbed preparation and because of this, no-till is often associated with increased chemical use in comparison to traditional tillage based methods of crop production. However, there are many agroecological alternatives to increased chemical use, such as winter cover crops and the mulch cover they provide, soil solarization or burning.

Management

No-till farming requires some different skills than conventional farming. As with any production system, if done incorrectly, yields can drop. A combination of technique, equipment, pesticides, crop rotation, fertilization, and irrigation have to be used for local conditions.

Cover Crops

In no-till occasionally uses cover crops to help control weeds and increase nutrients in the soil (by using legumes) or by using plants with long roots to pull mobile nutrients to the surface from lower layers of the soil. Cover crops then need to be killed so that the newly planted crops can get enough light, water, nutrients, etc. This can be done by rollers, crimper, choppers and other ways.

Crop Rotation

With no-till farming, residue from the previous years crops lie on the surface of the field, cooling it and increasing the moisture. This can increase or decrease disease or cause it to vary compared to tillage farming. To reduce weeds, pests and disease, crop rotation is used. Planting different crops year after year denies a pest or pathogen population a supply of whatever food it is adapted to consume.

Organic No-till

Some farmers who practice organic management often place ordinary, non-dyed corrugated cardboard on seed-beds and vegetable areas. Used correctly, cardboard placed on a specific area can:

1. keep important fungal hyphae and microorganisms in the soil intact,

2. prevent recurring weeds from popping up,

3. increase residual nitrogen and plant nutrients by top-composting plant residues,

4. create valuable topsoil for next year's seeds or transplants.

Plant residues (left over plant matter originating from cover crops, grass clippings, original plant life etc.) rots while underneath the cardboard so long as it remains moist enough. This rotting attracts worms and other beneficial microorganisms to the site of decomposition, and over a few seasons (usually Spring to Fall or Fall to Spring) and up to a few years, creates a layer of rich topsoil. Plants can then be seeded into the soil in spring, or holes can be cut into the cardboard to allow transplanting. Using this method in conjunction with other sustainable practices such as composting/vermicompost, cover crops and rotations are often considered beneficial to both land and those who take from it.

On fields too large to manually apply a residue with a high carbon-to-nitrogen ratio, a cover crop may be used to produce a similar effect. The cover crop may be killed by mowing or by crimping the stalk, as with a roller/crimper. The residue is then planted through, and left as a mulch to retard weed growth and slowly release the nutrients contained therein. Cover crops typically must be crimped when they enter the flowering/pollination stage.

Water Issues

No-till farming dramatically reduces erosion in a field. While much less soil is displaced, any gullies that form get deeper each year instead of being smoothed out by regular plowing. This may be handled by creating sod drainways, waterways, permanent drainways, cover crops, etc.

A problem in some fields is water saturation in soils. Switching to no-till farming corrects drainage because of the qualities of soil under continuous no-till include a higher water infiltration rate.

Equipment

It is very important to have planting equipment that can properly penetrate through the residue, into the soil and prepare a good seedbed. No-till farming requires much less maximum power from farm tractors, so equipment (a tractor) can be smaller than under tilling. Using a smaller, lighter tractor has the added benefit of reducing compaction and fuel consumption.

Soil Temperature

Another problem of no-till farming is that in spring, the soil both warms and dries more slowly, which may delay planting. The slower warming is due to crop residue being a lighter color than the soil which would be exposed in conventional tillage, which then absorbs less solar energy. This can be managed by using row cleaners on a planter. Since the soil can be cooler, harvest can occur a few days later than a conventionally tilled field. A cooler soil is also a benefit because water doesn't evaporate as fast.

Residue

On some crops, like continuous no-till corn, the thickness of the residue on the surface of the field can become a problem without proper preparation and equipment.

Fertilizer

One of the most common yield reducers is nitrogen being immobilized in the crop residue, which can take a few months to several years to decompose, depending on the crop's C to N ratio and the local environment. Fertilizer needs to be applied at a higher rate during the transition period while the soil rebuilds its organic matter. The nutrients in the organic matter are eventually released into the soil, so this is only a concern during the transition time frame (4–5 years for Kansas, USA). An innovative solution to this problem is to integrate animal husbandry in various ways to aid in decomposition.

Ridge tillage is a reduced tillage system where the crop is grown on top of permanent ridges. This system works well for fields that are often too wet to work in the spring. To initially construct ridges, start in the fall with a tilled field, and broadcast a cover crop that will winter-kill, like field peas and oats. Immediately construct the ridges and roll them to flatten the tops. In the spring, use a flail mower to chop the dead cover crop residue followed by wavy coulters or a rotary hoe to loosen the top inch of soil. This scrapes away the old crop residue and flattens the top of the ridge in order to plant the new crop. The ridge is then restored to full height during the final cultivation. Usually two cultivations are required to help control weeds, loosen the soil and re-construct the ridges. Straw can also be used between ridges to suppress weeds. The ridges can be replanted for many seasons before they need to be reconstructed. As with many reduced-till systems, specialized equipment is required for planting, cultivating and possibly harvesting. Ridge tillage helps conserve moisture, lower inputs, and provide a warmer and dryer soil environment for seeds.

Permanent bed systems help limit soil compaction and maintain soil structure. Equipment and foot traffic is limited to paths or tracks between the beds. Some permanent beds are raised structures while others are not. There are many different ways to construct permanent beds. One simple method is to use a spader to till the soil and provide a rotation between cover crops and cash crops to provide organic matter, nutrients, weed suppression and a great soil environment for healthy crops. Mulch is often used with permanent raised beds to add organic matter and suppress weeds.

For example, you can use a perennial sod cover crop for wheel tracks to avoid compaction on the beds and to increase habitat for beneficial insects. Properly prepared weed-free compost can be used to fertilize and simultaneously mulch the beds for weeds. Organic growers have found that constructing raised-beds, and then using tarps or a thick layer of weed-free compost as a mulch, reduces weed seeds over time, and the same beds can be used for years, with straw mulch to control weeds between beds.

Agroforestry

Agroforestry is an intensive land management system that optimizes the benefits from the biological interactions created when trees and shrubs are deliberately combined with crops and/or livestock. There are five basic types of agroforestry practices today windbreaks, alley cropping, silvopasture, riparian buffers and forest farming. Within each agroforestry practice, there is a continuum of options available to landowners depending on their own goals (e.g., whether to maximize the production of interplanted crops, animal forage, or trees).

Benefits of Agroforestry

The benefits created by agroforestry practices are both economic and environmental. Agroforestry can increase farm profitability in several ways:

1. The total output per unit area of tree/crop/livestock combinations is greater than any single component alone.

2. Crops and livestock protected from the damaging effects of wind are more productive.

3. New products add to the financial diversity and flexibility of the farming enterprise.

Agroforestry helps to conserve and protect natural resources by, for example, mitigating non-point source pollution, controlling soil erosion, and creating wildlife habitat. The benefits of agroforestry add up to a substantial improvement of the economic and resource sustainability of agriculture.

Key Traits of Agroforestry Practices

Agroforestry practices are intentional combinations of trees with crops and/or livestock which involve intensive management of the interactions between the components as an integrated agroecosystem. These four key characteristics - intentional, intensive, interactive and integrated - are the essence of agroforestry and are what distinguish it from other farming or forestry practices. To be called agroforestry, a land use practice must satisfy all of the following four criteria:

1. Intentional: Combinations of trees, crops and animals are intentionally de-signed and managed as a whole unit, rather than as individual elements which may occur in close proximity but are controlled separately.

2. Intensive: Agroforestry practices are intensively managed to maintain their productive and protective functions, and often involve annual operations such as cultivation, fertilization and irrigation.

3. Interactive: Agroforestry management seeks to actively manipulate the biolog-ical and physical interactions between the tree, crop and animal components. The goal is to enhance the production of more than one harvestable component at a time, while also providing conservation benefits such as non-point source water pollution control or wildlife habitat.

4. Integrated: The tree, crop and animal components are structurally and func-tionally combined into a single, integrated management unit. Integration may be horizontal or vertical, and above- or-ground. Such integration utilizes more of the productive capacity of the land and helps to balance economic production with resource conservation.

Agroforestry Practices

The five recognized agroforestry practices are:

1. Riparian and Upland Forest Buffers

Riparian forest buffers are strips of permanent vegetation, consisting of trees, shrubs, and grasses, planted or managed between agricultural land (usually crop-land or pastureland) and water bodies (rivers, streams, creeks, lakes, wetlands) to reduce runoff and non-point source pollution. Forest buffers are usually planted in three distinct zones near an agricultural stream for stabilizing streambanks, im-proving aquatic and terrestrial habitats, and providing harvestable products. Up-land buffers with cool- or warm-season grass alone or combined with shrubs and trees are also used to reduce nonpoint-source pollution and prevent gully formation in agricultural watersheds.

2. Windbreaks

Windbreak practices (shelterbelts, timberbelts, hedgerows, and living snowfences) are planted and managed as part of a crop or livestock operation to enhance crop produc-tion, protect crops and livestock, manage snow distribution, and/or control soil erosion.

Field windbreaks are used to protect a variety of wind-sensitive row crops, forage, tree, and vine crops to control soil erosion, and to provide other benefits such as improved insect pollination of crops and enhanced wildlife habitat.

Livestock windbreaks help reduce animal stress and mortality, improve feed and water consumption, enhance weight gain and calving success rates, and control odor. Timberbelts are managed windbreaks designed to increase the value of the forestry component.

3. Alley Cropping

This practice combines trees planted in single or multiple rows with agricultural or horticultural crops cultivated in the wide alleys between the tree rows. High-value hardwoods such as oak, walnut, ash, and pecan are favored species in alley cropping practices, and can potentially provide high-value lumber or veneer logs in the long-term.

Crops or forages grown in the alleys, and nuts from walnut, pecan and chestnut trees, pro- vide annual income from the land while the longer-term wood crop matures. Specialty crops (herbs, fruits, vegetables, nursery stock, flowers, etc.) can be grown in alleys, utilizing the microclimate created by trees, to boost economic production from each acre.

4. Silvopasture

This practice combines trees with forage (pasture or hay) and livestock production. Silvopasture can be established by adding trees to existing pasture, or by thinning an existing forest stand and adding (or improving) a forage component. Trees are managed for high-value timber or sawlogs, and at the same time they provide shelter for livestock, reduce heat stress and improve food and water consumption.

In the winter, the protection of trees reduces cold stress — therefore, animals do not lose as much energy keeping warm and are able to gain more weight. Forage and livestock provide short-term income at the same time a crop of high-value sawlogs is being grown, providing a greater overall economic return from the land.

5. Forest Farming

In forest farming practices, high-value specialty crops are cultivated under the protection of a forest overstory that has been modified and managed for sustained timber production and to provide the appropriate microclimate conditions.

Shade-tolerant specialty crops like ginseng, shiitake mushrooms, and decorative ferns grown in the understory are sold for medicinal/botanical, decorative/handicraft, or food products. Overstory trees are managed to produce timber and veneer logs.

A key concern in developing agroforestry nomenclature for the U.S. is overlap and confusion with mainstream land use management disciplines, e.g., forestry, agriculture, and livestock production. There is a fundamental need to develop a definition and criteria that would effectively distinguish practices that are agroforestry from those that are not. Application of the four criteria defining agroforestry (intentional, intensive, integrative, and interactive) provide the key to determine what is and is not an agroforestry practice.

Components of an Agroforestry System

Land

Agroforestry is not a system of pots on a balcony or in a greenhouse. It is a system by which land is managed for the benefit of the landowner, environment and long-term welfare of society. While appropriate for all landholdings, this is especially important in the case of hillside farming where agriculture may lead to rapid loss of soil. If the farmer owns the land, she/he has a vested interest in thinking conservatively, how the land can be maintained over long periods of time. Unfortunately, farmers who rent land may have less interest in the long-term benefits of agroforestry and may even fear that making improvements will raise the rent or result in the lease being terminated.

Trees

In agroforestry, particular attention is placed on multiple purpose trees or perennial shrubs. The most important of these trees are the legumes because of their ability to fix nitrogen and thus make it available to other plants. The roles of trees on the small farm may include the following:

- Sources of fruits, nuts, edible leaves, and other food.

- Sources of construction material, posts, lumber, branches for use as wattle (a fabrication of poles interwoven with slender branches etc.) and thatching.

- Sources of non-edible materials including sap, resins, tannins, insecticides, and medicinal compounds.

- Sources of fuel.

- Beautification.

- Shade.

- Soil conservation, especially on hillsides.

- Improvement of soil fertility.

In order to plan for the use of trees in agroforestry systems, considerable knowledge of their properties is necessary. Desirable information for each species includes its benefits, adaptability to local conditions (climate, soil, and stresses), the size and form of the canopy and root system, and suitability for various agroforestry practices. Some of the most common uses of trees in agroforestry systems are:

- Individual trees in home gardens, around houses, paths, and public places.

- Dispersed trees in cropland and pastures.

- Rows of trees with crops between (alley cropping).

- Strips of vegetation along contours or waterways.

- Living fences and borderlines, boundaries.

- Windbreaks.

- Improved fallows.

- Terraces on hills.

- Small earthworks.

- Erosion control on hillsides, gullies, channels.

- Woodlots for the production of fuel and timber.

Non-trees

Any crop plant can be used in agroforestry systems. The choice of crop plants in designing such systems should be based on those crops already produced in a particular region either for marketing, feeding animals, or for home consumption, or that have great promise for production in the region. In keeping with the philosophy of agroforestry, however, other values to be considered in crop selection include proper nutrition, self-sufficiency and soil protection. Thus, selection of crops requires a judgment based on knowledge of the crops, adaptations, production uses, as well as family needs, opportunities for barter, and markets.

Any farm animal can be used in agroforestry systems. The choice of animal will be based on the value the farmer places on animal-derived benefits including income, food, labor, non-food products, use of crop residues, and manure.

Drip Irrigation

Drip irrigation is a method of controlled irrigation in which water is slowly delivered to the root system of multiple plants. In this method water is either dripped onto the soil surface above the roots, or directly to the root zone. It is often a method chosen over surface irrigation because it helps to reduce water evaporation.

Drip irrigation is delivered to plant roots through a series of pipes, tubes, and valves. These parts, controlled by emitters and pumps, allow water to be focused in a particular area. In addition, drip systems can incorporate liquid fertilizer into the irrigation water.

Drip irrigation systems can help reduce evaporation and runoff, and contribute to water conservation. However, before this system can work correctly it must be properly installed and managed.

The two main types of drip irrigation are:

1. Surface drip irrigation - The water is delivered to the surface of the soil directly above the root system of the plants. This particular type of drip irrigation is mainly used on high-value crops.

2. Subsurface drip irrigation - The water is applied directly to the root system. This type is used particularly in growing row crops.

Drip System Layout

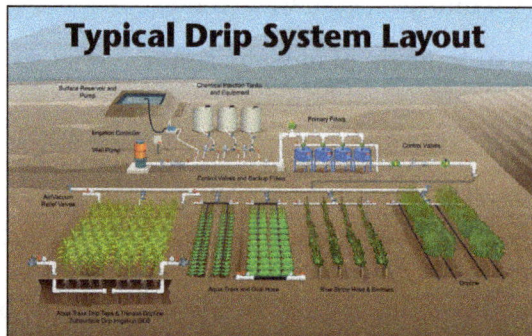

Layout of Drip Irrigation System

Major Components of Drip Irrigation System			
1	Pump station.	2	By-pass assembly
3	Control valves	4	Filtration system
5	Fertilizer tank /Venturi	6	Pressure gauge
7	Mains / Sub-mains	8	Laterals
9	Emitting devices	10	Micro tubes

Pump station takes water from the source and provides the right pressure for delivery into the pipe system.

Control valves control the discharge and pressure in the entire system.

Filtration system cleans the water. Common types of filter include screen filters and graded sand filters which remove fine material suspended in the water.

Fertilizer tank/venturi slowly add a measured dose of fertilizer into the water during irrigation. This is one of the major advantages of drip irrigation over other methods.

Mainlines, submains and laterals supply water from the control head into the fields. They are usually made from PVC or polyethylene hose and should be buried ground because they easily degrade when exposed to direct solar radiation. Lateral pipes are usually 13-32 mm diameter.

Emitters or drippers are devices used to control the discharge of water from the lateral to the plants. They are usually spaced more than 1 metre apart with one or more emitters used for a single plant such as a tree. For row crops more closely spaced emitters may be used to wet a strip of soil. Many different emitter designs have been produced in recent years. The basis of design is to produce an emitter which will provide a specified constant discharge which does not vary much with pressure changes, and does not block easily.

Wetting Pattern in Drip Irrigation

Unlike surface and sprinkler irrigation, drip irrigation only wets part of the soil root zone. This may be as, low as 30% of the volume of soil wetted by the other methods. The wetting patterns which develop from dripping water onto the soil depend on discharge and soil type. Figure shows the effect of changes in discharge on two different soil types, namely sand and clay.

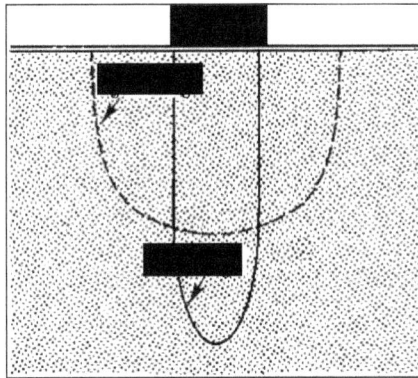

Wetting Pattern in Sandy Soils

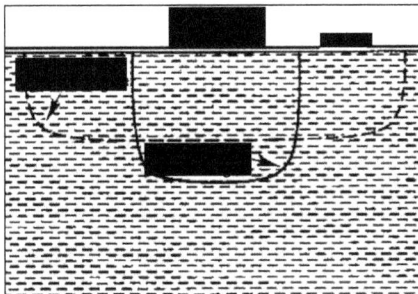

Wetting pattern in Clay Soils

Although only part of the root zone is wetted it is still important to meet the full water needs of the crop. It is sometimes thought that drip irrigation saves water by reducing the amount used by the crop. This is not true. Crop water use is not changed by the method of applying water. Crops just require the right amount for good growth.

The water savings that can be made using drip irrigation are the reductions in deep percolation, in surface runoff and in evaporation from the soil. These savings, it must be remembered, depend as much on the user of the equipment as on the equipment itself.

Drip irrigation is not a substitute for other proven methods of irrigation. It is just another way of applying water. It is best suited to areas where water quality is marginal, land is steeply sloping or undulating and of poor quality, where water or labour are expensive, or where high value crops require frequent water applications.

Crops Suitable for Drip Irrigation System

1.	Orchard Crops	Grapes, Banana, Pomegranate, Orange, Citrus, Mango, Lemon, Custard Apple, Sapota, Guava, Pineapple, Coconut, Cashewnut, Papaya, Aonla, Litchi, Watermelon, Muskmelon etc.
2.	Vegetables	Tomato, Chilly, Capsicum, Cabbage, Cauliflower, Onion, Okra, Brinjal, Bitter Gourd, Ridge Gourd, Cucumber, Peas, Spinach, Pumpkin etc.
3.	Cash Crops	Sugarcane, Cotton. Arecanut, Strawberry etc.
4.	Flowers	Rose, Carnation, Gerbera, Anthurium, Orchids, Jasmine, Dahilia, Marigold etc.
5.	Plantation	Tea, Rubber, Coffee, Coconut etc.
6.	Spices	Turmeric, Cloves, Mint etc.
7.	Oil Seed	Sunflower, Oil palm, Groundnut etc.
8.	Forest Crops	Teakwood, Bamboo etc.

Response of Different Crops to Drip Irrigation System

Crops	Water saving (%)	Increase in yield (%)
Banana	45	52
Cauliflower	68	70
Chilly	68	28
Cucumber	56	48
Grapes	48	23
Ground nut	40	152
Pomegranate	45	45
Sugarcane	50	99
Sweet lime	61	50
Tomato	42	60
Watermelon	66	19

Benefits of Drip Irrigation

- Increase in yield up to 230 %.

- Saves water up to 70% compare to flood irrigation. More land can be irrigated with the water thus saved.

- Crop grows consistently, healthier and matures fast.

- Early maturity results in higher and faster returns on investment.

- Fertilizer use efficiency increases by 30%.

- Cost of fertilizers, inter-culturing and labour use gets reduced.

- Fertilizer and Chemical Treatment can be given through Micro Irrigation System itself.

- Undulating terrains, Saline, Water logged, Sandy & Hilly lands can also be brought under productive cultivation.

Water Conservation Through Drip

Water is conserved in the following ways:

- Drip irrigation application uniformity is very high, usually over 90%.

- Unlike sprinklers, drip irrigation applies water directly to the soil, eliminating water loss from wind.

- Application rates are low so water may be spoon fed to the crop or plant root zone in the exact amounts required (even on a daily or hourly basis). In contrast, other methods entail higher water application quantities and less frequency. If young plants need water frequently, much of the water applied is often wasted to deep percolation or runoff.

- Low application rates are less likely to run off from heavier soils or sloping terrain.

- Drip irrigation does not water non-targeted areas such as furrows and roads in agriculture, between beds, blocks or benches in greenhouses, or hardscape, buildings or roads in landscape.

- Drip irrigation easily adapts to odd-shaped planting areas which are difficult to address with sprinklers or gravity irrigation.

- Drip irrigation is capable of germinating seeds and setting transplants which eliminates the need for "sprinklering up" and eliminates the resulting waste in the early stages of crop growth.

Drip irrigation is today's need because Water - nature's gift to mankind is not unlimited and free forever. World water resources are fast diminishing.

Green Manuring

Green manuring is defined as the growing of green manure crops & then turning off these crops directly in the field by ploughing the field so as to make the field richer in nitrogen which is the most deficient nutrient of the soil. Green manuring crops help in improving the structure of soil & also increases its physical properties. One of the main objectives of the green manuring is to increase the content of nitrogen in the soil to increase the crop production. The green manuring can be practised as in-situ or green leaf manuring. Green in-situ manuring refers to the growing of green manuring crops in the border rows or as intercrops along with the main crops for example: Sunn hemp, Cowpea, Dhaincha, Berseem, Green gram etc. whereas green leaf manuring is the collection of green leaves from outside places such as waste or forestlands; for example collection of wild dhaincha leaves & then incorporating them into the crop field for improving the soil properties. Most of the green manuring crops are incorporated in the fields after 6 to 8 weeks of sowing with the application of water so that they should be easily incorporated into the soil. The flowering stage of the green manuring crops is the best time for incorporation of these crops.

Procedure of Green Manuring

The green manuring practices (techniques) are given below:

1. Green manure crop can be grown in any type of soil, provided there is sufficient rainfall or alternatively irrigation available.

2. To ensure success with a leguminous green manure crop is to inoculate the seed with the proper strain of bacteria.

3. The green manure crop should be sown with a higher seed rate than usual so that there will be a good canopy produced very quickly. The usual seed rate for sannhemp is about 40 to 50 kg per hectare.

4. The production of green manures is limited by the plant food elements (plant nutrients) deficient in the soil. Leguminous green manure plants are able to fix atmospheric nitrogen. When the soil is rich in nitrogen, leguminous plants do not fix nitrogen so well, as when grown in poor soils. The application of phosphatic fertilizers improves the growth of leguminous crop markedly and promotes the fixation of nitrogen by profuse nodulation.

5. The best stage at which the crop should be incorporated in the soil as a green

manure is when it reaches the flowering stage. Sannhemp crop is ready for turning in at the age of 7 to 8 weeks whereas dhaincha crop is ready for incorporation when 5 to 6 weeks old.

6. Burying of green manure crop is done in the different ways. In some case the plants are cut close to the ground and the green material is put in the furrows opened by a mould board plough, and is later buried. One of the method is to plank the material down with a heavy plank or leg, and then plough the field. The other method is to mix the uprooted or cut plant material (green leaf manure) by means of disc harrow. In drier areas this method has been proved to be better than ploughing in.

7. Immediately after ploughing the material, careful packing of the soil should be done by suitable implements to ensure proper decomposition. Packing (compacting) is especially necessary if the soil moisture supply is deficient.

8. Under certain favourable circumstances, green manure crop such as dhaincha can be sown in between the rows of cotton or Jowar. When the dhaincha is sufficiently tall it can be uprooted and mixed with the soil by inter-cultivations.

9. Under limited moisture supply condition, it may be advisable to grow the green manure crops in one field and add the green material to another field. By doing this, the moisture required for growing the green manure crop is saved.

10. For proper decomposition, in light soils the crop should be buried deeper than that in the heavy ones.

Selection of Green Manure Crops

The characteristics of good green manuring crops are given below:

1. It should be quick growing, so that timely incorporation of green manure crops may be done. For example, sannhemp and mung.

2. It should yield large quantities of green material in a short period. For example, dhaincha.

3. It should be preferably from the legume family so that nitrogen would be fixed in addition to green matter production. For example, dhaincha and sannhemp.

4. It should be tender (move leafy growth than woody growth) so that its decomposition will be rapid. For example, berseem and lobia.

5. It should have a deep root system so that it would penetrate deep layers of the soil. Thus, it utilizes nutrients and water from deeper layers and also helps in developing good soil structure. For example, sannhemp and dhaincha.

Principles of Green Manuring:

- Green manure crop should be grown in irrigated area or where annual rainfall is more than 30 inches. Lack of moisture is harmful for the growth of the crop as well as for decomposition. An un-decomposed crop may harm the subsequent crop by upsetting the balance of carbon and nitrogen.

- After green manuring subsequent crops should be sown in well decomposed crops. Un-decomposed green manure may cause poor germination, and problem of diseases and insects.

- In irrigated area, the best stage at which the crop should be incorporated in the soil as a green manure is when it reaches the flowering stage. In rainfed or dry region, green manure crop should be incorporated before flowering stage (tender or leafy stage).

- Green manure crop should be quick growing.

- It should be preferably from the legume family.

- It should have deep root system.

Table: Green manure crops, their yield and nitrogen added in the field.

Name of Green manure crop	Growing season	Average yield of green matter (kg per hectare)	Nitrogen added (kg per hectare)
Crotalaria Juncea	Summer and Kharif	194.7	84.2
Sesbania aculeata	-do-	183.6	76.9
Phaseolus mungo	-do-	100.1	42.2
Phaseolus aureus	-do-	37.4	38.6
Lathyrus sativus	Rabi	123.0	54.9
Trifolium alexandrinum	-do-	155.0	54.2

Advantages of Green Manuring

Green manuring has the following advantages:

- Green manuring adds organic matter and nitrogen to the soil. Fresh organic matter (leaves, twigs, roots etc.) decomposes and liberates plant nutrients. Leguminous green manure crop fixes nitrogen in the soil.

- Green manuring checks weed growth. The plants used for green manuring are usually, grow very quickly and thus, tend to suppress the growth of weeds.

- Green manuring crops aid in the reclamation of saline and alkaline soils by the release of organic acids.

- Green manuring increases the availability of plant nutrients. When fresh organic matter decomposes, carbon dioxide is evolved, more organic acids are formed and as a. result, plant nutrients become more soluble in organic acids and therefore, more readily available to crops.

Disadvantages of Green Manuring

The limitations of green manuring are as follows:

- Growing green manure crops in rainfed area where annual rainfall is less than 30 inches, there is a likelihood of harmful effects. The reason for this is because the green material added to the soil does not decompose readily due to lack of sufficient water. Retarded decomposition results in nitrogen starvation of the following crop.

- Due to improper decomposition, problems of insect-pests and diseases may come up.

- Sometime the cost of green manuring is more than chemical fertilizers.

- Green manure crop may be taken as a catch crop between the main crops. Due to late sowing of green manure crop and insufficient moisture, burying of green manure crop become late. Therefore, sowing of main crop may not be done or delayed.

- If rainfall is scanty, growth of green manure crops would be less vigorous which results in less production of green matter.

Multiple Cropping

Multiple cropping is a form of Ecological Intensification that is potentially highly sustainable when two or more crops are grown at the same time or in a sequence. It does this by balancing three key ecological processes: competition, on the one hand, and commensalism (one plant gaining benefits from the other) or mutualism (both plants benefitting each other) on the other. Typically, farmers will plant crops as close together as possible to utilise all the available land. When different crop species or varieties are grown together, the competition may be fierce; trees grown in a maize field, for example, may shade out the crop. But this can be compensated for by determining the optimal spacing and by exploiting various forms of commensalism or mutualism, for example where the tree may be a legume, providing nitrogen for the crop plant beneath.

There are numerous examples of multiple cropping:

- Intercropping: Interspersion of different crops on the same piece of land, such as a home garden, either at random or more commonly in alternate rows usually designed to minimise competition but maximise the potential for both crops to make use of the available nutrients, such as nitrogen supplied by a legume.

- Rotations: The growing of two or more crops in sequence on the same piece of land.

- Agroforestry: Annual herbaceous crops are grown interspersed with perennial trees or shrubs. The deeper-rooted trees can often exploit water and nutrients otherwise unavailable to the crops. The trees may also provide shade and mulch, creating a microenvironment, whilst the ground cover of crops reduces weeds and prevents erosion.

- Sylvo-pasture: Similar to agroforestry, but combines trees with grassland and other fodder species for livestock grazing. The mixture of shrubs, grass and crops often supports mixed livestock populations.

- Green manuring: The growing of legumes and other plants to fix nitrogen and then incorporate the nutrients into the soil for the following crop. Commonly used green manures are Sesbania and the fern Azolla, which contains nitrogen-fixing, blue-green algae in ricefields.

Advantages of Multiple Cropping

- Increase the overall income. When crops are grown individually, individual crops may give better yield. But when crops are grown together individual yields of crops reduces but total yield are higher.

- Risk of growing one crop may overcome.

- Weed infestation become less.

- Insects and diseases infestation become less.

- Proper utilization of fertilizers. In our country about 25% nitrogen is utilized by crops, 75% are lost by evaporation and leaching. Growing of different crops together this problem can beovercome a little extent. When legume crops are growing together the crops may be beneficial.

- Advantages of having different food.

- It increases the intensity of cropping.

- Due to intensive cropping the poor farmers can increase their income.

- Development of self reliance.

- Employment of manpower/labor becomes wider.

- Expansion of industrial and marketing system.

Disadvantages of Multiple Cropping

- It may cause some problem to mechanization.
- Not effective in seed production.
- It may cause some problem when crops are not compatible.
- Climate may be favorable.
- Infestation of insects and diseases may occur.
- Intercultural operation is complex.
- Fertility reduces.

Sustainable Land Management

Land provides an environment for agricultural production, but it also is an essential condition for improved environmental management, including source/sink functions for greenhouse gasses, recycling of nutrients, amelioration and filtering of pollutants, and transmission and purification of water as part of the hydrologic cycle. The objective of sustainable land management (SLM) is to harmonise the complimentary goals of providing environmental, economic, and social opportunities for the benefit of present and future generations, while maintaining and enhancing the quality of the land (soil, water and air) resource. Sustainable land management is the use of land to meet changing human needs (agriculture, forestry, conservation), while ensuring long-term socioeconomic and ecological functions of the land.

Sustainable land management is a necessary building block for sustainable agricultural development. Sustainable agricultural development, conservation of natural resources, and promoting sustainable land management are key objectives of the new World Bank rural investment program, From Vision to Action, and increasingly these objectives are being included in all agricultural development and natural resources management projects.

Sustainable land management combines technologies, policies, and activities aimed at integrating socioeconomic principles with environmental concerns, so as to simultaneously:

- Maintain and enhance production (productivity).
- Reduce the level of production risk, and enhance soil capacity to buffer against degradation processes (stability/resilience).
- Protect the potential of natural resources and prevent degradation of soil and water quality (protection).

- Be economically viable (viability).

- Be socially acceptable, and assure access to the benefits from improved land management (acceptability/equity).

The definition and these criteria, called pillars of SLM, are the basic principles and the foundation on which sustainable land management is being developed. Any evaluation of the sustainability has to be based on these objectives: productivity, stability/resilience, protection, viability, and acceptability/equity. The definition and pillars have been field tested in several countries, and they were judged to provide useful guidance to assess sustainability.

The lack of a comprehensive, quantifiable definition for sustainable land management is sometimes considered to be a serious deficiency. Yet, as argued by Gallopin, a research model for sustainability has to be more flexible and therefore less easy to quantify than a research model for chemistry, physics, or classical agronomy. Such a research model must be designed around an evaluation process (rather than thematic context), because it is intended to test the likelihood of certain events taking place and the aggregate impacts of these events, rather than specifics of various null hypothesis or the impacts of certain inputs or land management interventions. Essentially the research model must include a goal statement, a conceptual framework, a set of procedures, and criteria (indicators) for diagnosis. One of the main objectives of such a research model is to evaluate the impacts of events which are uncertain, but the process of evaluation is guided by scientifically defined protocols.

Principles and Criteria for Sustainable Land Management

Experiences gained from field projects in developing and developed countries has identified a series of principles for sustainable land management, and these can be used as general guidelines for development projects. The most useful of these are summarised:

Global Concerns for Sustainability

- Sustainability can be achieved only through the collective efforts of those immediately responsible for managing resources. This requires a policy environment that empowers farmers and other, local decision makers, to reap benefits for good land use decisions, but also to be held responsible for inappropriate land uses.

- Integration of economic and environmental interests in a comprehensive manner is necessary to achieve the objectives of sustainable land management. This requires that environmental concerns be given equal importance to economic performance in evaluating the impacts of development projects, and that reliable indicators of environmental performance be developed.

- There is urgent need to resolve the global challenge to produce more food to feed rapidly rising global populations, while at the same time preserving the biological production potential, resilience, and environmental maintenance systems of the land. Sustainable land management, if properly designed and implemented, will ensure that agriculture becomes a part of the environmental solution, rather than remaining an environmental problem.

Sustainable Agriculture

- More ecologically balanced land management can achieve both economic and environmental benefits, and this must be the foundation (linch pin) for further rural interventions (investments). Without good land management, other investments in the rural sector are likely to be disappointing2. At the same time, arguing for the continued maintenance of agriculture without reference to environmental sustainability is increasingly difficult. Indicators of land quality are needed to guide us along the way.

- Agricultural intensification is often necessary to achieve more sustainable systems. This requires shifts to higher value production, or higher yields with more inputs per unit of production and higher standards of management (more knowledge intensive). However, sustainable agriculture has to work within the bounds of nature not against them. Many yield improvements can be achieved by optimizing efficiency of external inputs rather than trying to maximize yields.

- The importance of off-farm income should not be underestimated because it i) supplements cash flow on the farm, ii) generates an investment environment for improved land management, and therefore iii) reduce production pressures on land.

Sharing Responsibilities for Sustainability

- Farmers and land managers must expand their knowledge of sustainable technologies and implement improved procedures of land stewardship. The preferred option is not to tell the farmer what to do (command and control legislation), but to create an enabling environment through policy interventions where farmers are more free to make the right choice. A policy environment where farmers are more empowered, but also held accountable, for achieving the objectives of sustainable land management is essential. However, sustainable land management is the responsibility of all segments of society. Governments must ensure that their policies and programs do not create negative environmental impacts, and society needs to define requirements for land maintenance and develop a "social" discount rate for future land use options that encourages the most sustainable use.

- Concerns for sustainable land management go beyond agriculture to include the legitimate interests of other aspects of land stewardship, including wildlife, waterfowl and biodiversity management. There is increasing evidence that society is demanding that farmers become stewards of rural landscapes, and that agriculture become more than simply putting food on the table. Many of society's environmental values may not represent economic gains for farmers, however, and farmers cannot shoulder all the costs of environmental maintenance.

Relationships Among Soil Quality, Land Quality and Sustainable Land Management

New concepts of soil and land quality are emerging, and often these are used interchangeably. These concepts and their relationships are summarized, to the extent that some concensus is available on how these should be applied.

Soil quality is the capacity of a specific soil to function within natural or managed ecosystem boundaries to sustain plant and animal production, maintain or enhance water quality, and support human health and habitation.

Land quality is the condition, state or "health" of the land relative to human requirements, including agricultural production, forestry, conservation, and environmental management.

Sustainable land management combines technologies, policies, and activities aimed at integrating socio-economic principles with environmental concerns so as to simultaneously maintain or enhance production, reduce the level of production risk, protect the potential of natural resources and prevent (buffer against) soil and water degradation, be economically viable, and be socially acceptable.

These concepts span the scales of detail, application, and levels of integration with socio-economic data. Soil quality is the most restrictive, followed by land quality and then sustainable land management. Soil quality is effectively a condition of a site, and it can be studied using soil data alone. Land quality requires integration of soil data with other biophysical information, such as climate, geology and land use. Land quality is a condition of the landscape, i.e. it is a biophysical property, but includes the impacts of human interventions (land use) on the landscape. Sustainable land management requires the integration of these biophysical conditions, i.e. land quality, with economic and social demands. It is an assessment of the impacts of human habitation, and a condition of sustainable development.

These are more than simple differences in semantics; the concepts differ in the kinds and scale of the processes being described, the data used for input, and the amount and kinds of integration with other disciplines. However, the concepts form a continuum over the landscape, and they must be applied for different types and scales of land use.

Land Quality Indicator Program

Assessment of sustainable land management requires appropriate evaluation instruments, such as Land Quality Indicators. However, land quality, like the concept of sustainable land management of which it is a part, requires operational definitions and specific, measureable indicators if it is to be more than an attractive, conceptual phrase.

The World Bank, in collaboration with UNEP, UNDP, FAO and the CGIAR, is developing a program called Land Quality Indicators (LQIs), as a means to better coordinate actions on land related issues such as land degradation. In the area of economic and social data, and in some cases for air and water quality, indicators are already in regular use to support decision-making at global, national and sub-national levels. In contrast, few such indicators are available to assess, monitor and evaluate changes in the quality of land resources. Land refers not just to soil but to the combined resources of terrain, water, soil and biotic resources that provide the basis for land use. Land quality refers to the condition or "health" of land, and specifically to its capacity for sustainable land use and environmental management.

The LQI program addresses the dual objectives of environmental monitoring as well as sector performance monitoring for managed ecosystems (agriculture, forestry, conservation and environmental management). It is being developed for application at national and regional scales, but it is also part of a larger, global effort on improved natural resources management. The LQI program is in response to the United Nations Conference on Environment and Development, and it fits with Agenda 21 expectations as well as the Convention to Combat Desertification.

Agricultural cropland, including agroforestry, as well as forested lands, range and pasture lands, are under increasing pressure because of population migrations to marginal land areas and agricultural intensification on existing cultivated lands. Sustainable land use intensification requires the maintenance or enhancement of the productive potential of the land resources, i.e. increases in food supplies must come from agricultural intensification rather than from area expansion, but this must be done without degrading the land resource on which production depends. The question is how this should be achieved, and how to monitor progress towards this objective in the different agroecological regions of developing countries. LQIs are tools to help us along the way.

In general and particularly in developing countries, it is essential that scarce resources devoted to land management be used more cost-efficiently and that policy-makers have at least rough indicators of whether environmental conditions and land quality are getting better or worse. Land quality indicators, such as nutrient balance, loss of organic matter, land use intensity and diversity, and land cover are useful to task managers and decision makers to monitor and improve the performance of projects with respect to their socio-economic and environmental impacts, and to assess the trend towards or away from land use sustainability. While routine project performance indicators based on cost-benefit analyses (input-output factors, risk and economic performance indicators), are necessary to monitor the activities and components of a project, LQIs are required to evaluate the environmental impact(s). The quantitative assessment of physical impacts, such as depletion of soil nutrients, loss of organic matter, soil erosion, water contamination etc. may appear to be costly and cumbersome during project implementation, but the long-term negative impact of reduced land quality, such as decreased efficiency of fertilizers, increased erosion, increased fuel consumption, increased pest infestation (nematodes, etc.), often result in rehabilitation costs that are much higher. The LQI approach focuses on preventive maintenance rather than rehabilitation, and provides the methodology and the approach to integrate the socio-economic and biophysical information that are required for better informed sustainable land management strategies.

The development of LQIs follows a logical framework, providing information not only on the state of the resources, but also the underlying causes (or "pressure") as well as the response of the society to the state and the pressure exerted on the land resources. Because of the nature and the complexity of land issues, the LQI program recommends addressing issues of land management by agro-ecological zones (Resource Management Domains or "Terroirs"). This new approach puts the focus on evaluating impacts of human interventions on specific landscapes, rather than emphasizing only biophysical variables as was the case in the past. At the same time, this spatial stratification favors incorporating farmer and other local knowledge into the overall process of improved agricultural and environmental land management.

A research strategy for the Land Quality Program was developed during a two day research planning meeting, Washington, D.C., 1996, sponsored by the World Bank. A panel of internationally acclaimed scientists and administrators established the objectives and priorities for the research, defined strategic alliances to be developed with ongoing national and international programs, and identified potential sources of funding. They also achieved international agreement on a core set of strategic land quality indicators.

Core LQIs for managed ecosystems (agriculture and forestry) in the major agro-ecological zones (AEZs) of tropical, sub-tropical and temperate environments, and recommended for development in the short term include:

- Nutrient balance: This describes nutrient stocks and flows as related to different land management systems used by farmers in specific AEZs and specific countries.

- Yield trends and yield gaps: This describes current yields, yield trends, and actual:potential, farm level yields for the major food crops in different countries.

- Land use intensity: Describes the impacts of agricultural intensification on land quality. Intensification may involve increased cropping, more value-added production, and increased amounts and frequency of inputs; emphasis is on the management practices adopted by farmers in the transition to intensification.

- Land use diversity (agro-diversity): Describes the degree of diversification of production systems over the landscape, including livestock and agro-forestry systems; it reflects the degree of flexibility (and resilience) of regional farming systems, and their capacity to absorb shocks and respond to opportunities.

- Land cover: Describes the extent, duration and timing of vegetative cover on the land during major erosive periods of the year. It is a surrogate for erosion, and along with land use intensity and diversity, it will give increased understanding on the issues of desertification.

A second set of core LQIs were recommended for longer-term research. These are indicators which require further development of their theoretical base, or lack adequate data for immediate development. These include:

- Soil Quality: Likely to be based on soil organic matter turn-over, particularly the dynamic (microbiological) carbon pool most affected by environmental conditions and land use change.

- Land degradation (erosion, salinization, compaction, organic matter loss): These processes have been much researched and have a strong scientific base, but reliable data on extent and impacts are often lacking.

- Agro-biodiversity: Involves objectives of managing natural habitats and the co-existence of native species in agricultural areas, maintaining natural soil micro and meso biodiversity, and managing the gene pools utilized in crop and animal production.

Four additional sets of core LQIs were identified, but these were recommended to be developed through collaboration with the respective authoritative disciplines:

- Water quality

- Forest land quality

- Rangeland quality

- Land contamination/pollution.

The above are the biophysical components of sustainable land management. Although useful in their own right, they must still be complemented with indicators of the other pillars of sustainable land management, economic viability, system resilience, and social equity and acceptability.

Shifting Cultivation

Shifting agriculture is a system of cultivation that preserves soil fertility by plot (field) rotation, as distinct from crop rotation. In shifting agriculture a plot of land is cleared and cultivated for a short period of time; then it is abandoned and allowed to revert to its natural vegetation while the cultivator moves on to another plot. The period of cultivation is usually terminated when the soil shows signs of exhaustion or, more commonly, when the field is overrun by weeds. The length of time that a field is cultivated is usually shorter than the period over which the land is allowed to regenerate by lying fallow.

One land-clearing system of shifting agriculture is the slash-and-burn method, which leaves only stumps and large trees in the field after the standing vegetation has been cut down and burned, its ashes enriching the soil. Cultivation of the earth after clearing is usually accomplished by hoe or digging stick and not by plow.

Shifting agriculture has frequently been attacked in principle because it degrades the fertility of forestlands of tropical regions. Nevertheless, shifting agriculture is an adaptation to tropical soil conditions in regions where long-term, continued cultivation of the same field, without advanced techniques of soil conservation and the use of fertilizers, would be extremely detrimental to the fertility of the land. In such environments it may be preferable to cultivate a field for a short period and then abandon it before the soil is completely exhausted of nutrients.

Fallow fields are not unproductive. During the fallow period, shifting cultivators use the successive vegetation species widely for timber for fencing and construction, firewood, thatching, ropes, clothing, tools, carrying devices and medicines. It is common for fruit and nut trees to be planted in fallow fields to the extent that parts of some fallows are in fact orchards. Soil-enhancing shrub or tree species may be planted or protected from slashing or burning in fallows. Many of these species have been shown to fix nitrogen. Fallows commonly contain plants that attract birds and animals and are important for hunting. But perhaps most importantly, tree fallows protect soil against physical erosion and draw nutrients to the surface from deep in the soil profile.

The relationship between the time the land is cultivated and the time it is fallowed are critical to the stability of shifting cultivation systems. These parameters determine whether or not the shifting cultivation system as a whole suffers a net loss of nutrients over time. A system in which there is a net loss of nutrients with each cycle will eventually lead to a degradation of resources unless actions are taken to arrest the losses. In some cases soil can be irreversibly exhausted (including erosion as well as nutrient loss) in less than a decade.

The longer a field is cropped, the greater the loss of soil organic matter, cation-exchange-capacity and in nitrogen and phosphorus, the greater the increase in acidity, the more likely soil porosity and infiltration capacity is reduced and the greater the loss of seeds of naturally occurring plant species from soil seed banks. In a stable shifting cultivation system, the fallow is long enough for the natural vegetation to recover to the state that it was in before it was cleared, and for the soil to recover to the condition it was in before cropping began. During fallow periods soil temperatures are lower, wind and water erosion is much reduced, nutrient cycling becomes closed again, nutrients are extracted from the subsoil, soil fauna decreases, acidity is reduced, soil structure, texture and moisture characteristics improve and seed banks are replenished.

The secondary forests created by shifting cultivation are commonly richer in plant and animal resources useful to humans than primary forests, even though they are much less bio-diverse. Shifting cultivators view the forest as an agricultural landscape of fields at various stages in a regular cycle. People unused to living in forests cannot see the fields for the trees. Rather they perceive an apparently chaotic landscape in which trees are cut and burned randomly and so they characterise shifting cultivation as ephemeral or 'pre-agricultural', as 'primitive' and as a stage to be progressed beyond. Shifting agriculture is none of these things. Stable shifting cultivation systems are highly variable, closely adapted to micro-environments and are carefully managed by farmers during both the cropping and fallow stages. Shifting cultivators may possess a highly developed knowledge and understanding of their local environments and of the crops and native plant species they exploit. Complex and highly adaptive land tenure systems sometimes exist under shifting cultivation. Introduced crops for food and as cash have been skillfully integrated into some shifting cultivation systems. Its disadvantages include the high initial cost, as manual labour is required.

Environmental Change and Societies

Shifting cultivation in Indonesia. A new crop is sprouting through the burnt soil.

A growing body of palynological evidence finds that simple human societies brought about extensive changes to their environments before the establishment of any sort of state, feudal or capitalist, and before the development of large scale mining, smelting or shipbuilding industries. In these societies agriculture was the driving force in the economy and shifting cultivation was the most common type of agriculture practiced. By examining the relationships between social and economic change and agricultural change in these societies, insights can be gained on contemporary social and economic change and global environment change, and the place of shifting cultivation in those relationship.

As early as 1930 questions about relationships between the rise and fall of the Mayan civilization of the Yucatán Peninsula and shifting cultivation were raised and continue to be debated today. Archaeological evidence suggests the development of Mayan society and economy began around 250 AD. A mere 700 years later it reached its apogee, by which time the population may have reached 2,000,000 people. There followed a precipitous decline that left the great cities and ceremonial centres vacant and overgrown with jungle vegetation. The causes of this decline are uncertain; but warfare and the exhaustion of agricultural land are commonly cited. More recent work suggests the Maya may have, in suitable places, developed irrigation systems and more intensive agricultural practices.

Similar paths appear to have been followed by Polynesian settlers in New Zealand and the Pacific Islands, who within 500 years of their arrival around 1100 AD turned substantial areas from forest into scrub and fern and in the process caused the elimination of numerous species of birds and animals. In the restricted environments of the Pacific islands, including Fiji and Hawaii, early extensive erosion and change of vegetation is presumed to have been caused by shifting cultivation on slopes. Soils washed from slopes were deposited in valley bottoms as a rich, swampy alluvium. These new environments were then exploited to develop intensive, irrigated fields. The change from shifting cultivation to intensive irrigated fields occurred in association with a rapid growth in population and the development of elaborate and highly stratified chiefdoms. In the larger, temperate latitude, islands of New Zealand the presumed course of events took a different path. There the stimulus for population growth was the hunting of large birds to extinction, during which time forests in drier areas were destroyed by burning, followed the development of intensive agriculture in favorable environments, based mainly on sweet potato (Ipomoea batatas) and a reliance on the gathering of two main wild plant species in less favorable environments. These changes, as in the smaller islands, were accompanied by population growth, the competition for the occupation of the best environments, complexity in social organization, and endemic warfare.

The record of humanly induced changes in environments is longer in New Guinea than in most places. Agricultural activities probably began 5,000 to 9,000 years ago. However, the most spectacular changes, in both societies and environments, are believed to have occurred in the central highlands of the island within the last 1,000 years, in association with the introduction of a crop new to New Guinea, the sweet potato. One of the most striking signals of the relatively recent intensification of agriculture is the sudden increase in sedimentation rates in small lakes.

The root question posed by these and the numerous other examples that could be cited of simple societies that have intensified their agricultural systems in association with increases in population and social complexity is not whether or how shifting cultivation was responsible for the extensive changes to landscapes and environments. Rather it is why simple societies of shifting cultivators in the tropical forest of Yucatán, or the highlands of New Guinea, began to grow in numbers and to develop stratified and sometimes complex social hierarchies?

At first sight, the greatest stimulus to the intensification of a shifting cultivation system is a growth in population. If no other changes occur within the system, for each extra person to be fed from the system, a small extra amount of land must be cultivated. The total amount of land available is the land being presently cropped and all of the land in fallow. If the area occupied by the system is not expanded into previously unused land, then either the cropping period must be extended or the fallow period shortened.

At least two problems exist with the population growth hypothesis. First, population growth in most pre-industrial shifting cultivator societies has been shown to be very low over the long term. Second, no human societies are known where people work only to eat. People engage in social relations with each other and agricultural produce is used in the conduct of these relationships.

These relationships are the focus of two attempts to understand the nexus between human societies and their environments, one an explanation of a particular situation and the other a general exploration of the problem.

Feedback Loops

In a study of the Duna in the Southern Highlands of New Guinea, a group in the process of moving from shifting cultivation into permanent field agriculture post sweet potato, Modjeska argued for the development of two "self amplifying feed back loops" of ecological and social causation. The trigger to the changes were very slow population growth and the slow expansion of agriculture to meet the demands of this growth. This set in motion the first feedback loop, the "use-value" loop. As more forest was cleared there was a decline in wild food resources and protein produced from hunting, which was substituted for by an increase in domestic pig raising. An increase in domestic pigs required a further expansion in agriculture. The greater protein available from the larger number of pigs increased human fertility and survival rates and resulted in faster population growth.

The outcome of the operation of the two loops, one bringing about ecological change and the other social and economic change, is an expanding and intensifying agricultural system, the conversion of forest to grassland, a population growing at an increasing rate and expanding geographically and a society that is increasing in complexity and stratification.

Resources are Cultural Appraisals

The second attempt to explain the relationships between simple agricultural societies and their environments is that of Ellen. Ellen does not attempt to separate use-values from social production. He argues that almost all of the materials required by humans to live (with perhaps the exception of air) are obtained through social relations of production and that these relations proliferate and are modified in numerous ways. The values that humans attribute to items produced from the environment arise out of cultural arrangements and not from the objects themselves, a restatement of Carl Sauer's dictum that "resources are cultural appraisals". Humans frequently translate actual objects into culturally conceived forms, an example being the translation by the Duna of the pig into an item of compensation and redemption. As a result, two fundamental processes underlie the ecology of human social systems: First, the obtaining of materials from the environment and their alteration and circulation through social relations, and second, giving the material a value which will affect how important it is to obtain it, circulate it or alter it. Environmental pressures are thus mediated through social relations.

Transitions in ecological systems and in social systems do not proceed at the same rate. The rate of phylogenetic change is determined mainly by natural selection and partly by human interference and adaptation, such as for example, the domestication of a wild species. Humans however have the ability to learn and to communicate their knowledge to each other and across generations. If most social systems have the tendency to increase in complexity they will, sooner or later, come into conflict with, or into "contradiction" with their environments. What happens around the point of "contradiction" will determine the extent of the environmental degradation that will occur. Of particular importance is the ability of the society to change, to invent or to innovate technologically and sociologically, in order to overcome the "contradiction" without incurring continuing environmental degradation, or social disintegration.

An economic study of what occurs at the points of conflict with specific reference to shifting cultivation is that of Esther Boserup. Boserup argues that low intensity farming, extensive shifting cultivation for example, has lower labor costs than more intensive farming systems. This assertion remains controversial. She also argues that given a choice, a human group will always choose the technique which has the lowest absolute labor cost rather than the highest yield. But at the point of conflict, yields will have become unsatisfactory. Boserup argues, contra Malthus, that rather than population always overwhelming resources, that humans will invent a new agricultural technique or adopt an existing innovation that will boost yields and that is adapted to the new environmental conditions created by the degradation which has occurred already, even though they will pay for the increases in higher labor costs. Examples of such changes are the adoption of new higher yielding crops, the exchanging of a digging stick for a hoe, or a hoe for a plough, or the development of irrigation systems. The controversy over Boserup's proposal is in part over whether intensive systems are more costly in labor terms, and whether humans will bring about change in their agricultural systems before environmental degradation forces them to.

Contemporary Shifting Cultivation Practice

Sumatra, Indonesia

Rio Xingu, Brazil

Santa Cruz, Bolivia

Kasempa, Zambia

The estimated rate of deforestation in Southeast Asia in 1990 was 34,000 km² per year. In Indonesia alone it was estimated 13,100 km² per year were being lost, 3,680 km² per year from Sumatra and 3,770 km² from Kalimantan, of which 1,440 km² were due to the fires of 1982 to 1983. Since those estimates were made huge fires have ravaged Indonesian forests during the 1997 to 1998 El Niño associated drought.

Shifting cultivation was assessed by the FAO to be one of the causes of deforestation while logging was not. The apparent discrimination against shifting cultivators caused a confrontation between FAO and environmental groups, who saw the FAO supporting commercial logging interests against the rights of indigenous people. Other independent studies of the problem note that despite lack of government control over forests and the dominance of a political elite in the logging industry, the causes of deforestation are more complex. The loggers have provided paid employment to former subsistence farmers. One of the outcomes of cash incomes has been rapid population growth among indigenous groups of former shifting cultivators that has placed pressure on their traditional long fallow farming systems. Many farmers have taken advantage of the improved road access to urban areas by planting cash crops, such as rubber or pepper as noted above. Increased cash incomes often are spent on chain saws, which have enabled larger areas to be cleared for cultivation. Fallow periods have been reduced and cropping periods extended. Serious poverty elsewhere in the country has brought thousands of land-hungry settlers into the cut-over forests along the logging roads. The settlers practice what appears to be shifting cultivation but which is in fact a one-cycle slash and burn followed by continuous cropping, with no intention to long fallow. Clearing of trees and the permanent cultivation of fragile soils in a tropical environment with little attempt to replace lost nutrients may cause rapid degradation of the fragile soils.

The loss of forest in Indonesia, Thailand, and the Philippines during the 1990s was preceded by major ecosystem disruptions in Vietnam, Laos and Cambodia in the 1970s and 1980s caused by warfare. Forests were sprayed with defoliants, thousands of rural forest dwelling people uproots from their homes and moved and roads driven into previously isolated areas. The loss of the tropical forests of Southeast Asia is the particular outcome of the general possible outcomes described by Ellen when small local ecological and social systems become part of larger system. When the previous relatively stable ecological relationships are destabilized, degradation can occur rapidly. Similar descriptions of the loss of forest and destruction of fragile ecosystems could be provided from the Amazon Basin, by large scale state sponsored colonization forest land or from the Central Africa where what endemic armed conflict is destabilizing rural settlement and farming communities on a massive scale.

Zero Waste Agriculture

Zero waste agriculture (ZWA) is a type of sustainable agriculture which optimises use of the five natural kingdoms that is plants, animals, bacteria, fungi and algae to produce bio diverse food, energy and nutrients in a synergistic integrated cycle of profit making process where the waste of each process becomes the feed stock for another process. Zero waste agriculture preserves local indigenous systems and existing agrarian cultural values and practices.

Sustainable agriculture integrates three main goals, environmental health, economic profitability and social and economic equity. In Haritha Organic Farm"s sustainable project, all the waste materials on the plantation is decomposed and put back in the soil as manure. Weeds and other biomass from the plantation are finely chopped and mixed with cow dung and water and put back in the soil as manure. For Integrated farming process we provides the technology for biogas digester the heart of the Zero waste Agriculture system.

Advantages

1. Optimise food production in an ecological sound manner.

2. Reduces water consumption through and recycling and reduced evaporation.

3. Provides energy security through the harvesting of bio methane (biogas).

4. Provides climate change relief through the substantial reduction in green house gas emissions from both traditional agriculture practice and fossil fuel usage.

5. Reduces the use of pesticides through natural remedies and bio-pesticides.

6. Reduce the use of synthetic fertilisers through the use of composting techniques such as vermi composting, green manuring etc.

Naturescaping

Naturescaping, also called natural landscaping, is an idea that supports local biodiversity, rather than some sort of static, externally-defined concept of what a yard "should" look like.

In naturescaping, gardeners use local plants, locally-grown, to design and fill their gardens. They work with the contours of the land rather than against them. Traditional gardens, in contrast, adapt the land to the plant, often with a heady mix of fertilizers and other chemicals.

Traditional gardens are largely the creation of landscape architects trained in specific ways, coupled with a culturally-induced expectation of what a yard should look like. The green lawns of Arizona are a classic example of this, requiring watering several times a weeks in the middle of a desert. In this type of gardening, stock solutions are sought, rather than best solutions. While these solutions are often beautiful, they don't always fit the land, and often do not make use of native materials or sources.

Naturescaping has a number of benefits, both for the home-owner and for the environment. Naturescaping saves both time and money. A garden planted with local plants rarely needs to be watered once it established, and is low-maintenance. Since it requires no pesticides, herbicides, or fertilizers, it is less expensive to keep, and far healthier for both the environment and the homeowner.

Between 30% and 60% of a household's water bill goes to watering the lawn. This alone should indicate a problem. A natural landscape needs little or no extra water, as the plants are adapted to the local rain conditions.

These local plants are also better suited to attracting native birds and small wildlife. Loss of habitat is contributing to the annual 5%-10% decline in songbird populations, but naturescaping works to start rebuilding lost ground.

A natural yard is a more dynamic place, less-planned, with more possibilities to let the imagination wander. While some view this as a benefit, others view it as a drawback.

The natural landscape is much more random than a planned, picked-from-a-garden-centre sort of place. A local tree may not conform to our expectations, and while is kind of the point, for some it causes difficulty. It is more difficult to create a specific "look" with this brand of gardening, and letting nature take its course may also appear sloppy. This can be an issue in some areas, especially areas where strata agreements or Home Owner Associations (HOAs) have rules in place governing yard appearance. Even where such rules do not exist, naturescaping can draw sharp looks from neighbors, and even the occasional call from local officials concerned about weeds. However, properly maintained, it should not be a problem.

As concerns about chemical and fertilizer use continue to grow, the natural landscape is becoming more and more the logical, practical, and proper choice for a yard. It is now a matter of educating people to the benefits, and beauty, of the naturescape.

Benefits

There are many benefits to naturescaping, whether practiced in place of or in addition to traditional landscaping. The benefits include, but are not limited to, the following which are expanded upon on our Benefits Page:

- Low Maintenance - Native plants evolved to grow in local conditions and to predictable sizes. They do not require watering (except during establishment), chemical pesticides and fertilizers, or frequent cutting.

- Public Health (lowers cancer rates) - Traditional landscaping uses large amounts of synthetic pesticides and fertilizers, some of which are suspected carcinogens. During rains, these chemicals often run off into public water supplies. Traditional landscaping also contributes to air and noise pollution.

- Saves you Money - The cost of maintaining a naturescape is dramatically less than that of a traditional landscape because a naturescape essentially takes care of itself. Naturescapes also save you time - and how valuable is your time?

- Water - In the West, 60% of consumed water goes to lawns; in the East, 30%. This water diversion harms the environment, kills fish, and returns polluted water to our streams and rivers. It also costs you - on irrigation system installation and maintenance, and on your water bill.

- Song Birds - Our song bird populations having dropped steadily - 5-10%, per year!, depending on the species - for the last several decades, and there is no end in sight. The loss is primarily due to habitat loss. Adopting naturescaping is critical if song birds are to remain.

- Enhances Livability - An ecologically functional landscape offers so much more than a sterile, static landscape. It offers imagination to our children, and color, sound and wonder to all of us. It is cleaner, quieter and healthier, and may increase property values.

No Dig Gardening

The no-dig gardening concept was popularised by Sydney gardener Esther Dean in the 1970s as a way of minimising gardening effort while kick-starting a garden with maximum fertility. Any more fertility and you're likely to have triplets. A no-dig garden consists of layers of organic material that are stacked up to form a rich, raised garden area. The no-dig garden can be whatever height you desire. Vegetable seedlings, flowering annuals, herbs, bulbs and strawberries all thrive in a no-dig garden.

Benefits of No Dig Gardening

- This type of garden can be set up anywhere – over a lawn, inside a box frame, or even over concrete.

- No-dig gardens are quick and easy to make.

- If your soil is not ideal for veggie growing, a no-dig garden creates a great soil mix to plant into.

- No-dig gardens are very fertile as the decomposing organic matter quickly becomes rich, black compost and attracts beneficial micro-organisms.

- It retains moisture well.

- It discourages the growth of weeds as the soil is not turned over (burying weed seeds in moist soil).

Materials

- Newspaper (let's turn bad news into good).

- A decent amount of blood-and-bone or chicken manure (if building garden over grass or weeds).

- A largish amount of bulk manure – e.g., horse, cow, sheep or wildebeest – think grazing animal.

- High carbon (usually brown) organic material – e.g., straw, autumn leaves or dry grass clipping.

- Compost (black, rich, broken down organic matter).

The Technique of No-Dig Gardening

There are many variations of how we can build a "no dig gardening", but they all use the same underlying principle, which is soil building. No-dig gardens can be constructed anywhere because this technique creates soil – a rich, dark, healthy, nutrient-filled humus which plants love. They can be constructed over soil, existing lawn or concrete.

As a brief description, the way the technique of no-dig gardening works is that of different organic materials such as pea straw, lucerne, animal manure, finely-chopped prunings, kitchen scraps, compost and laid down in layers over each other to create what is essentially a thick, flat composting system that fills a garden bed. To plant seedlings or plants into such a garden bed, small 'pockets' or holes are made that hold as much compost as a small pot that you could grow the plant in, they are then filled with compost, and the plants planted into them. It's really simple, and the results are incredible. Essentially the no-dig garden is constructed of alternating layers of carbon-rich and nitrogen-rich materials, just like a properly made compost heap.

This diagram shows how a no-dig garden bed is typically built:

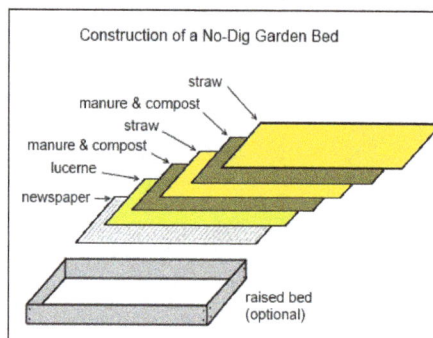

Construction of a No-Dig Garden Bed

straw
manure & compost
straw
manure & compost
lucerne
newspaper

raised bed (optional)

Build a No-Dig Garden in Ten Easy Steps

Building a no-dig garden is a very simple technique that doesn't take very long. I teach no-dig gardening classes where students get to build a no-dig garden for the first time ever, and a small group can easily construct and fully plant up a 1m x 4m (3' × 12') no-dig garden in around 30 minutes.

There are two main construction methods for building a no-dig garden:

- No dig gardens built on concrete, paved areas or rocky ground.

- No dig gardens built on existing garden beds or lawns.

The only difference is that you need to add an extra layer first when building on hard or rocky surfaces.

Here are the step-by step instructions for building a no-dig garden:

Step 1 – Select and Mark Location

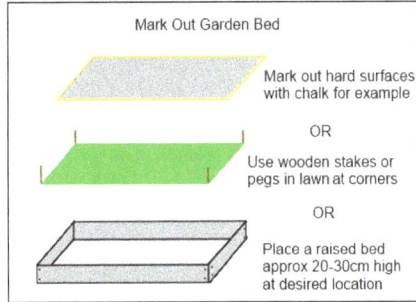

Mark Out Garden Bed

Mark out hard surfaces with chalk for example

OR

Use wooden stakes or pegs in lawn at corners

OR

Place a raised bed approx 20-30cm high at desired location

- Select a suitable location to construct a no-dig garden bed. Ideally it should be on a fairly level surface, and it should receive 5 hours or more of sunlight each day.

  You can build the no-dig garden over any surface, over existing soil, lawn, concrete or paved surfaces – the first step of the construction will vary depending on the surface.

- Either mark out where the no-dig garden bed will be, and build it without 'sides 'or edging, or construct a raised bed.

Step 2 – Gather Materials

You will need the following materials:

- Newspapers or cardboard

- Animal manure or organic fertilizer

- Straw bales or lucerne (alfalfa hay) bales or both

- Compost.

Optional materials:

- Kitchen scraps, worm castings, rock dust.

If building on hard or rocky ground, you'll also need:

- dry small sticks and branches, old dry leaves

- dry seaweed (optional).

You will also need the following items:

- If using cardboard – Bucket of water for soaking cardboard

- Watering can or hose for watering.

Step 3 – Preparing the Ground

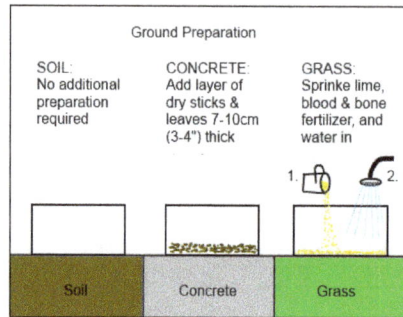

- If building over an existing garden bed or soil, no additional preparation is required.

- If building over concrete, paving, rocky ground or other hard surfaces, first lay down a layer of small sticks and branches, twigs and old dry leaves 7-10cm (3"-4") thick.

- This layer helps with drainage so water doesn't pool on the hard surface and create a waterlogged soil.You can also add dried seaweed (if you can get it) to this layer.

- If building over lawn or grass, you can mow the grass very low first, or just leave it. Next, fertilise it with plenty of nitrogen-rich fertiliser (such as blood & bone or manure) and lime, then water it in. The fertiliser will help the grass rot down once it is covered up and buried under all the layers that will go on top of it.

Step 4 – Lay Down Newspaper

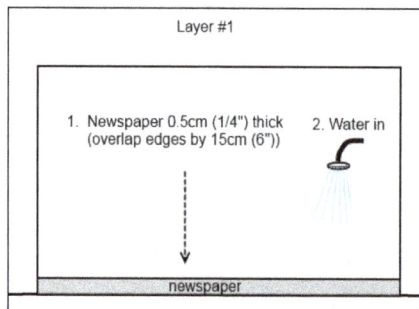

- Lay down sheets of newspaper in layers approximately 0.5cm thick (approx.

1/4" thick), and overlap the edges by 10-15cm to prevent grass or weeds growing through.

- Using a watering can or hose, water the newspaper well.

This newspaper layer will hold moisture and act as a weed barrier. It will gradually break down over time.

If using cardboard, you will need to pre-soak it in a bucket of water first, which is not as easy. The other issue with cardboard is that it contains glue made of borax, so it's really a second choice.

Use newspapers if they are available, and more importantly, do not use glossy printed paper or office paper, they contain toxic inks and bleaches, something you don't want going into your food.

Step 5 – Lay Down Lucerne

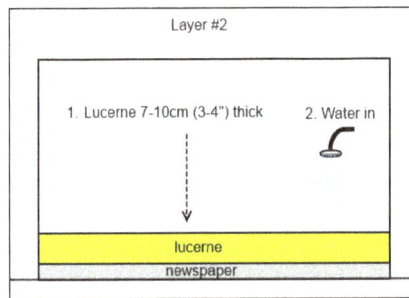

- Lay down a layer of lucerne approximately 10cm (4") thick over the newspaper.

  Using a watering can or hose, water in well.

You can use any other carbon containing material such as peas straw, hay, sugar cane mulch, etc, but lucerne is preferable because it has a higher nitrogen content than the other straw materials, and breaks down more easily. The carbon to nitrogen ration (C:N) for lucerne (alfalfa hay) is 18:1, while the straw is 80:1.

Step 6 – Lay Down Manure and Compost

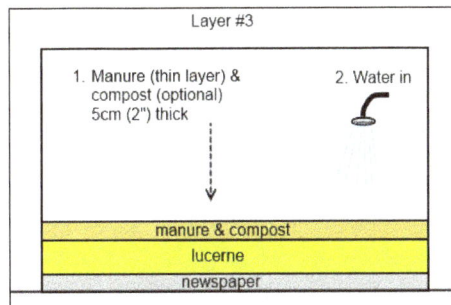

- Sprinkle a thin layer of manure. You can also add compost to create a layer 5cm (2") thick.

- Using a watering can or hose, water in well.

NOTE: If you want to add other ingredients such as kitchen scraps, worm castings, or rock dust into your no dig garden, this is the layer you add them to. Just use a thin layer, don't overdo it! The worm castings and rock dust can also be used in the upcoming higher layers, but kitchen scraps need to be placed in this lower layer only to keep it well buried, this prevents vermin such as rats and mice digging it up to get to it.

Step 7 – Lay Down Straw

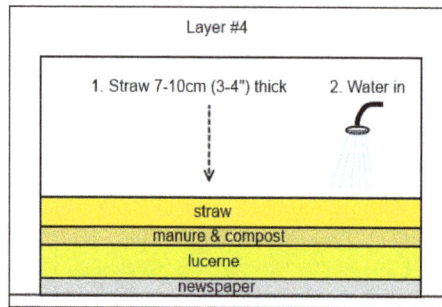

- Lay down a layer of straw approximately 10cm (4") thick over the layer of manure or manure/compost.

- Using a watering can or hose, water in well.

You can use any carbon containing material here such as peas straw, hay, sugar cane mulch, etc.

Step 8 – Lay down Manure and Compost

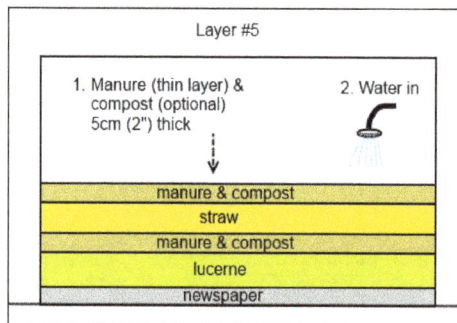

- Sprinkle a thin layer of manure. You can also add compost to create a layer 5cm (2") thick.

- Using a watering can or hose, water in well.

If you want to add other ingredients such as worm castings or rock dust into your no dig garden, you can also add them to this layer.

Step 9 – Lay Down Straw

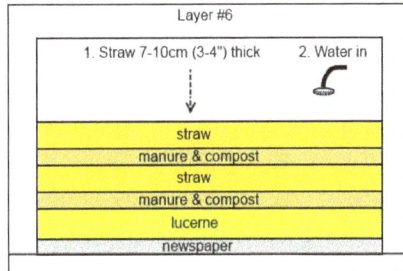

- Lay down another layer of straw approximately 10cm (4") thick over the layer of manure or manure/compost.

- Using a watering can or hose, water in well.

You can use any carbon containing material such as peas straw, hay, sugar cane mulch, etc here.

Step 10 – Make Pockets of Compost in Top Layer and Plant up

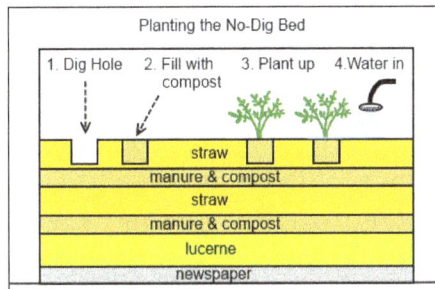

- Make holes in the top layer of straw approximately 10-15cm (4-6") wide, and equally deep.

- Fill with compost.

- Plant seeds, seedlings or plants.

- Using a watering can or hose, water in well.

You can also add seaweed extract to the water when you water in the seeds/seedlings or plants. Plants really do need more than the basic NPK (nitrogen, phosphorus, potassium) of chemical fertilizers. Seaweed contains just about every beneficial mineral, including all the trace elements that plants need, and it really helps your plants develop strong, healthy roots.

Now you've finished, just step back and admire your newly constructed no-dig garden bed! It's that easy, and that's how you build it, in 10 simple steps.

Many Approaches to No-Dig Gardening

The steps above outline just one of the many no-dig gardening 'formulas'. There are many ways to build no dig gardens, and there are many recipes for what to use for each layer. Some no-dig gardens can be very high and free-standing, while others can be low. They all work because they use the magic natural formula that we also use in composting, alternating layers of carbon-rich materials and nitrogen-rich materials.

So far we've only discussed building no-dig gardens as a means of creating new garden beds, but the beauty of this system is that you can also convert existing garden beds to a no-dig system, and it's even easier.

Converting Existing Gardens to No-Dig Gardens

Building new gardens from scratch is one thing, but it doesn't happen very often. More often we encounter a tired, run-down garden bed where the soil is depleted and compacted, where nothing much grows in it other than weeds. It's even easier to 'retrofit' and existing garden, to renovate it and convert it into a no-dig garden.

You can even use this technique to transform a fully planted garden bed into a no-dig garden, as you'll simply be laying down a two-layer mulch!

Converting an existing garden bed to a no-dig system involves three basic steps:

Prepare the Soil

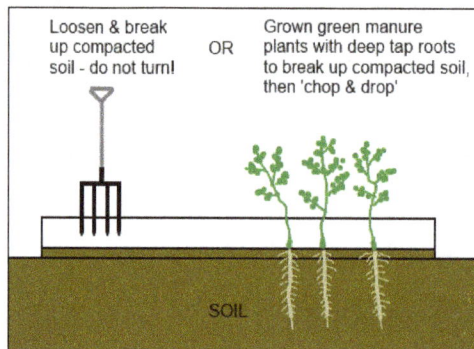

If necessary, loosen compacted soil. If the soil is not compacted. You can loosen compacted soil manually with a garden fork, which takes a few minutes, or you can plant 'green manure' plants with deep tap roots, which will drill into the compacted soil and break it up. This will take much longer (a few months) as the plant grows through its growing season. Once the plants starts to flower, it is cut down and the soil level and dropped on the soil surface, with the roots left in the ground – this is "chop & drop".

The roots will then break down and create deep air and water channels, and the soil will be loosened up naturally.

1. Cool Season 'green manure' plants which have deep taproots that can be used to break up compacted soil – Fenugreek, Lupins, Woolly Pod vetch.

2. Warm Season 'green manure' plants which have deep taproots that can be used to break up compacted soil – Lucerne.

The most important thing to do is to give an old garden bed a head start, once the soil is loosened, it will never be compacted ever again, because you don't step in a no-dig garden.

To loosen compacted soil, break it up with a garden fork, but don't turn it over – we're just trying to make the soil loose and friable here, we're not trying to kill all the soil ecology, which is what turning the soil does in conventional gardening, and why it's done.

Once we soil is loosened, and the no-dig garden layers are added, they will start to break down, adding organic matter to the soil, and the earthworms will do all the digging from there on, taking the nice organic matter from the surface and carrying it further into the soil, slowly converting the soil underneath into a rich, dark humus – real soil! remember, humans don't dig, earthworms do, and they do a much better job than us, so leave the digging to the experts, the earthworms, and save your time and energy.

Lay Down Manure and Compost

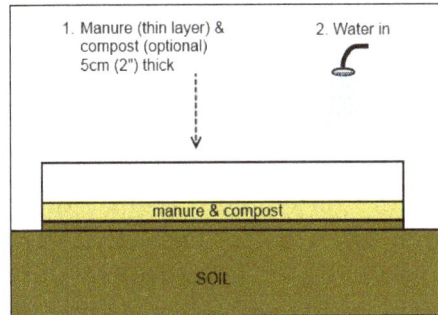

1. Sprinkle a thin layer of manure. You can also add compost to create a layer up 5cm (2") thick, but a thinner layer is just fine.

If there are existing plants in the garden bed, keep the materials away from the stems/ trunks to avoid 'collar rot' – rotting the base of the plant.

You can also add other ingredients such as worm castings (a rich fertilizer filled with lots of beneficial soil organisms) or rock dust (a slow release source of trace elements and minerals) into this layer.

2. Using a watering can or hose, water in well.

You can add seaweed extract to the water, it's rich in potassium which helps fruiting and flowering, and is loaded with lots of minerals which help the plants develop a strong and healthy root system.

Lay Down Straw

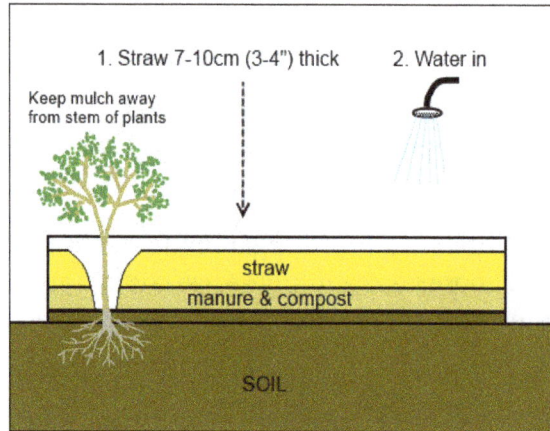

1. Lay down a layer of straw approximately 10cm (4") thick over the layer of manure or manure/compost.

If there are existing plants in the garden bed, keep the mulch away from the stems/trunks to avoid 'collar rot' – rotting the base of the plant.

2. Using a watering can or hose, water in well. You can use seaweed extract with the water once again.

You can use any carbon containing material here such as peas straw, hay, sugar cane mulch, etc.

Make sure you don't step in the garden and compact the soil once again, use paths – garden beds are for plants, paths are for humans Remember, you and the plants have conflicting needs, they need soft, loose friable soil that they can easily sink their roots into, you want firm stable paths you can walk across that you won't sink into.

No-Dig Garden Maintenance

To maintain any type of no-dig garden, you only need to repeat the above three steps detailed in "Converting Existing Gardens to No-Dig Gardens".

After the plants are harvested and the growing season has ended, the layers will have rotted down into the soil, enriching and improving it. It's then time to replenish the no-dig layers. You'll replenish them at the end of each major season (when all the winter crops are finished, and then again when all the summer crops are finished).

To replenish the layers of the no-dig garden:

1. Add a layer of manure as before (and compost if you wish, which is optional).

2. Cover the manure/compost layer with a layer of straw.

3. Water it in.

Controlled Environment Agriculture

Controlled environment agriculture (CEA) is the process of growing plants inside a greenhouse or grow room. The controlled environment allows the grower to maintain the proper light, carbon dioxide, temperature, humidity, water, pH levels, and nutrients to produce crops year-round.

In many cases, hydroponic growing systems are used for controlled environment agriculture to ensure that the plants receive optimal nutrients and water needed to produce an ample crop.

The entire process of controlled environment agriculture focuses on making the most of space, labor, water, energy, nutrients, and capital to operate while still producing a bountiful harvest.

Controlled environment agriculture allows a grower to reduce the incidences of pests or disease, increase overall efficiency, save resources, and even recycle things such as water or nutrients.

Using artificial lighting also increases crop production and allows many plants to be grown and produce year round by creating an optimum 12-month growing season.

Researchers frequently use controlled environment agriculture facilities to isolate specific plants and study their production in a maintained setting. In such an area all aspects, that affect the growth of a plant can be monitored so that precise data may be collected for scientific study.

The CEA cultivation process can be done in virtually any form of contained area, whether that be a skyscraper, home, or warehouse. This flexibility permits plants, and therefore food, to be grown in almost any location, creating agricultural opportunities in typically infertile areas, such as deserts, cities, or outer space. Moreover, by using this method food may be produced at any time of the year, because CEA's regulated environment is not subject to the same weather-constrained growing seasons that traditional agriculture is. Subsequently, as concerns build surrounding conventional field agriculture, including its impact on topsoil degradation, water usage, and distance from urban centers (to name just a few), CEA is increasingly being looked toward as a viable alternative for modern food production.

Working of Controlled Environment Agriculture

The most common factors controlled in CEA

This agricultural method functions by controlling for several factors that influence the growth rate and health of crops. These factors most often include:

1. Temperature (air, nutrient solution, root-zone)

2. Humidity

3. Carbon dioxide

4. Light (intensity, spectrum, interval)

5. Nutrient concentration

6. Nutrient pH (acidity).

In order to effectively manage growing conditions, CEA farmers typically engineer technological systems, which adjust the input and output of nutrients and resources to the plants within their enclosures. Over time, several forms of such technologies have been experimented with, and refined, to the point where they are now commonly applied within CEA operations. Some notable examples of CEA technologies are the frequently publicized methods of hydroponics, vertical farming, and LED light growing.

Benefits of Controlled Environment Agriculture

1. Much Lower Water Consumption:

   CEA systems are optimized to minimize evaporation and excess — using no more than the amount of water required by each crop.

2. Cleaner Growing Practices (For You and the Environment):

   A CEA system has many fewer pests, weeds, and diseases to contend with. This means that CEA growers do not need to place nasty pesticides and herbicides in your food and waterways, like traditional agriculture does.

3. Better Location and Distribution to Cities:

   Because a CEA system may be located almost anywhere, and crops can be grown using much less land, it is easier to position a farm within or close to urban centers.

4. No Usage of GMOs Necessary:

   The monitoring and mechanical solutions utilized in CEA systems ensure consistently high, healthy yields, making the usage of GMOs unnecessary.

5. Consistent Availability:

   A CEA system ensures that crops are always in-season and experience optimal climates, regardless of whether there may be frosts or droughts outside. This means that consumers get the freshest, most reliably grown produce all year round.

Organic Fertilizers

Organic fertilizers are the ones sourced from organic materials such as plants, mineral or animal sources. Unlike the traditional chemical fertilizers, organic fertilizers have to occur naturally. The organic fertilizers vary based on the nutrient requirements for the firm, but in most cases, the organic fertilizers are comprised of a single ingredient.

The nutritional and ingredient value of organic fertilizers doesn't provide immediate fix compared to the chemical fertilizers. Instead, they slowly break down by the action of organisms and biological processes in order for the plants to acquire the nutrients while at the same time conditioning and rejuvenating the soil.

Organic fertilizers are therefore eco-friendly and that's why they are preferred in organic and healthy farming. The chief examples of organic fertilizers include fish extracts, plant waste from agriculture, animal waste, treated sewage sludge, and peat.

Types of Organic Fertilizers

Dry

Exactly what it says on the tin, dry fertilizers are often mixed into the soil. They can be used on both in-ground gardens and container grown plants. These types of fertilizer are generally added as a way of encouraging long-term growth in seedlings, transplants, and crops.

Liquid

Obviously, these fertilizers are nutrients in liquid form. They may additionally use a type of binding agent to help them be better absorbed by the plant in question. These fertilizers might be poured onto the soil surrounding the plant so that they can be absorbed roots. Or they could be sprayed on the leaves instead. After all, foliar (leaf) sprays are particularly useful for vegetables during their growing season.

Liquid fertilizers are generally considered good for plants that are actively growing and should usually be applied on a monthly basis. However, leafy crops might need to be sprayed on a biweekly basis.

Growth Enhancers

While they aren't fertilizers per se, these substances help plants absorb nutrients that they receive from elsewhere. Some of them, such as kelp, are also a great source of trace elements. However, paying for these elements to be included in your fertilizer can be a waste of money since healthy soil should already contain helpful substances such as microbes, enzymes, and humic acids.

Using an Organic Fertilizer

You pretty much use organic fertilizers in the same way you would use regular chemical fertilizer. If you are buying premade supplies, the process is fairly simple. Read the instructions on the box and err on the side of caution when you're deciding how much to apply.

However, if you want to mix up your own fertilizer for a specific garden need that you have, you can certainly do that as well. Just be careful not to overdo it in order to avoid burning or even killing sensitive plants.

List of Organic Fertilizers

Alfalfa Meal

Alfalfa Meal

Nitrogen, Phosphorus and Potassium Ratio: 3-2-2

Alfalfa is commonly used as part of livestock feed. However, alfalfa meal is simply ground up so that it breaks down faster. This particular fertilizer has low amounts of nitrogen, phosphorus, and potassium. As a result, alfalfa meal works at moderate rate of speed. The best use for this fertilizer is as a soil conditioner in the early spring prior to planting crops.

Cottonseed Meal

Cottonseed Meal

- Nitrogen, Phosphorus and Potassium Ratio :6-1-1

This fertilizer has plenty of nitrogen, but it also contains fair amounts of phosphorus and potassium. The downsides to cottonseed meal are that it works slowly and that it is available primarily in cotton growing areas. However, this fertilizer is particularly useful for conditioning gardens in the fall before cover crops are planted or before mulch is applied. This gives the cotton seed meal time to break down fully so that the nitrogen present is readily available in the spring.

Corn Gluten Meal

Corn Gluten Meal

- Nitrogen, Phosphorus and Potassium Ratio :0.5-0.5-1

Corn gluten meal contains trace amounts of nitrogen, phosphorous, and potassium. It is also good soil stabilizer but it works slowly. Therefore, you should add it in the fall so that it has time to break down over the winter.

Rock Phosphate

Rock Phosphate

- Nitrogen, Phosphorus and Potassium Ratio: 0-5-0

This fertilizer is made from rocks that have been ground up. It contains large amounts of phosphate as well as other essential nutrients. The main benefit of using rock phosphate is that the elements it contains don't dissolve in water. Instead they hang around in the soil until they're used by the plants that are growing there.

Cow Manure

Cow Manure

- Nitrogen, Phosphorus and Potassium Ratio: 2.5-1-1.5

Manure in general has a high mass to nutrients ratio. It nonetheless contains respectable amounts of nitrogen, phosphorus, and potassium. Cow manure also works on gardens at a moderate rate of speed. These elements all help to make it an excellent compost additive. However, some weed seeds may survive being digested by the cows in question and this can cause obvious problems. You should also avoid manure leftover from industrial operations because it contains lots of salt. However, even regular manure can end up burning plants if too much is used or if it's used too often.

Chicken or Poultry Manure

Chicken Manure

- Nitrogen, Phosphorus and Potassium Ratio: 3.5-1.5-1.5

Having a lot of poultry crap on hand doesn't mean that you'll get comparatively high nutrient levels. Even so, this fertilizer contains slightly higher amounts of the three main plant nutrients than cow manure does. It's also works in a somewhat faster fashion. The best time to use poultry manure is just after harvesting your crops or just before you begin another gardening cycle. As is the case with cow manure, you'll want to be careful with this product because it can burn your plants if too much is applied.

Earthworm Castings

Earthworm Castings

- Nitrogen, Phosphorus and Potassium Ratio: 2-1-1

Earthworm castings contain decent amounts of all three vital nutrients. As a result, this type of fertilizer is considered a great addition to flower and vegetable gardens.

Greensand

Greensand

- Nitrogen, Phosphorus and Potassium Ratio: 1-1-5

Greensand comes from ancient sea beds. This high calcium fertilizer also contains iron, potassium, and other trace elements. However, the nutrient levels in greensand products can vary depending on their source.

Compost

Compost

- Nitrogen, Phosphorus and Potassium Ratio: 2-1.5-1.5

Compost's nutrient profile varies based on what is put into but it is often close to that of cow manure. It is a popular garden fertilizer. This is no doubt because it can be made for free and it works at a moderate pace. There are also a variety of ways that compost can be used in your garden. It can be used as a mulch or mixed with your garden soil. It can also be brewed into compost tea to use as a foliar feed. Gardeners who don't have space for a large compost system can even get similar benefits from using worm compost bins under their sinks.

However, compost can be alkaline in nature. This characteristic negatively affects how well plants living in the soil can absorb nutrients. Compost also has a high weight to available nutrient ratio. Improperly tended compost pile can additionally emit bad smells, which can quickly get city dwellers in trouble with their neighbors.

If you still plan on using it, compost should be added before or after planting. It's also a good soil refresher in between growing seasons for gardens that are constantly in use. You'll want to use half an inch to a full inch every time a new crop is planted.

Soybean Meal

Soybean Meal

- Nitrogen, Phosphorus and Potassium Ratio: 7-2-0

Soybean meal is high nitrogen fertilizer that also contains low amounts of phosphorus and calcium. Local growing conditions tend to affect how fast this product is absorbed into the soil but this process normally occurs at moderate rates of speed. However, soybean meal is fairly useful as a long term soil conditioner.

Blood Meal

Blood Meal

- Nitrogen, Phosphorus and Potassium Ratio : 12-1.5-0.5

This fertilizer is created from the powdered blood of butchered livestock. While it is high in nitrogen, blood meal is low in other elements. It is also highly acidic and likely to burn plants if too much is used at one time. You should definitely proceed with caution. It's best to apply blood meal to the soil before planting anything in it. Even so, blood meal's fast-acting nature makes it a good tonic for ailing plants. You will need to pour this fertilizer over the plants roots according to the package directions and then possibly cover everything up with mulch.

Bone Meal

Bone Meal

- Nitrogen, Phosphorus and Potassium Ratio: 4-20-0

It's no surprise that bone meal is made from ground up cow bones. This high phosphorous fertilizer also contains plenty of nitrogen. Bone meal works at moderate speeds to encourage flower production and root growth. As a result, it is great for flowering plants, bulbs, and fruit trees. However, bone meal is mostly used as a soil amendment for spots with high nitrogen levels where plants keep getting burned.

Feather Meal

Feather Meal

- Nitrogen, Phosphorus and Potassium Ratio: 12-0-0

Feather meal is very high in nitrogen. However, it doesn't contain any calcium or phosphorus and it only works at moderate speeds. If you plan on using this fertilizer in your yard, use it prior to planting in order to give the soil a nutrient boost.

Seabird Guano

Seabird Guano

- Nitrogen, Phosphorus and Potassium Ratio: 10-10-2

Seabird guano is gathered from islands that have low rainfall and arid climates, both factors which help it retain high levels of nutrients. Seabird guano contains large levels of trace elements along with decent amounts of nitrogen, phosphorus, and calcium. In fact, it is considered to be among the world's best organic fertilizers.

Bat Guano

Bat Guano

- Nitrogen, Phosphorus and Potassium Ratio: 10-10-2

Bat guano is a fast-acting fertilizer that contains a wide variety of nutrients such as nitrogen, phosphorus, calcium, and various other minerals. It is not as high in these components as seabird guano but it is more readily available. Since some of the nutrients contained in bat guano are water-soluble, this fertilizer is probably at its most effective when used as a foliar spray or a compost tea. You can also apply it between crop plantings and as a soil refresher in the late spring.

Fish Meal

Fish Meal

- Nitrogen, Phosphorus and Potassium Ratio: 5-2-2

Due to its high nitrogen levels, fish meal is a fairly fast acting fertilizer. It also has decent amounts of phosphorus and calcium. It is good for corn crops.

Fish Emulsion

Fish Emulsion

- Nitrogen, Phosphorus and Potassium Ratio: 2-4-0 to 5-1-1

This product is made from partially decomposed fish. As a result, it often has a bad odor. Fish emulsion fertilizer is high in nitrogen but it contains no calcium or potassium. It is also very acidic and should be used lightly to avoid burning plants. Fish emulsion nonetheless acts immediately once it is applied, which makes it a good treatment for leafy species that are suffering from low nitrogen levels. However, some plants may not tolerate it very well.

Shellfish Fertilizer/Shell Meal

Shellfish Meal

- Nitrogen, Phosphorus and Potassium Ratio: 5-2-5

This fertilizer is made from crushed up seafood byproducts. It contains plenty of calcium and some phosphorous as well as a large quantity of trace minerals. Shell meal also contains an element called chitin, which helps ward off pesky nematodes.

Liquid Kelp Fertilizer

Liquid Kelp Fertilizer

- Nitrogen, Phosphorus and Potassium Ratio: 1-0.2-2

This is a kelp based fertilizer created by cold processing. Kelp contains small amounts of the main three fertilizer components but it's quite high in trace elements. It is also a good source of growth hormones that can help plants reach their full potential. This liquid fertilizer is typically mixed with water and used as a foliar spray or poured into the soil around plants.

Seaweed

Seaweed

- Nitrogen, Phosphorus and Potassium Ratio: 1.5-0.75-5

Seaweed is a fast acting fertilizer that's often available for free along most coastlines. It contains all of the major three nutrients in small amounts but also contains plenty of zinc and iron. Seaweed is considered highly beneficial to grain crops as well as those that need high levels of potassium.

Grass Clippings

Grass Clippings

- Nitrogen, Phosphorus and Potassium Ratio:1-0-1.2

It's hard to beat grass clippings as a low cost fertilizer since they're usually free. This substance can be used to prevent weeds and conserve moisture in the soil. However, the nitrogen content in each batch varies. A layer of 1 to 2 inches should be sufficient for a full growing season. Just don't use cuttings from lawns that have been grown using herbicides.

References

- Principle-of-crop-rotation: agrihunt.com, Retrieved 11 May, 2019

- Introduction-to-cover-cropping-in-organic-farming-systems: extension.org, Retrieved 17 April, 2019

- Definition-of-cover-crop: thespruce.com, Retrieved 27 March, 2019

- Introduction-to-cover-cropping-in-organic-farming-systems: extension.org, Retrieved 23 January, 2019

- Reduced-tillage-no-tillage: satavic.org, Retrieved 3 April, 2019

- Reduced-tillage, cultural-practices: nevegetable.org, Retrieved 18 August, 2019

- What-is-agroforestry: aftaweb.org, Retrieved 8 May, 2019

- Agroforestry-Principles: umass.edu, Retrieved 25 June, 2019

- Drip-irrigation: maximumyield.com, Retrieved 5 March, 2019

- Drip-irrigation-system, farm-machinary, agriculture: vikaspedia.in, Retrieved 13 May, 2019

- Green-manuring: abcofagri.com, Retrieved 21 July, 2019

- Green-manuring-procedure-principles-and-advantages, organic-manures: soilmanagementindia.com, Retrieved 28 April, 2019

- Multiple-cropping, ecological-intensification: ag4impact.org, Retrieved 14 June, 2019

- Multiple-cropping: hasanuzzaman.weebly.com, Retrieved 17 August, 2019

- Shifting-agriculture: britannica.com, Retrieved 29 February, 2019

- Zero-waste-farming: harithaorganicfarms.com, Retrieved 30 June, 2019

- Naturescaping: greenerideal.com, Retrieved 14 January, 2019

- No-dig-gardening: veryediblegardens.com.au, Retrieved 9 July, 2019

- No-dig-gardening: deepgreenpermaculture.com, Retrieved 11 May, 2019

- Controlled-environment-agriculture: maximumyield.com, Retrieved 21 March, 2019

- Benefits-and-types-of-organic-fertilizers: conserve-energy-future.com, Retrieved 1 February, 2019

- Organic-fertilizers: epicgardening.com, Retrieved 27 January, 2019

Chapter 5

Sustainable Forest Management

The practice of managing the forest resources in such a way that it preserves the health of the forest while fulfilling the requirements of the society is known as sustainable forest management. Some of the practices which are involved in sustainable regulation of forests are coppicing, pollarding and forest gardening. The diverse aspects of sustainable forest management have been thoroughly discussed in this chapter.

Sustainable forest management, also known as sustainable forestry, is the practice of regulating forest resources to meet the needs of society and industry while preserving the forest's health. Therefore, sustainable forest management is always looking to strike a balance between the demand for the forest's natural resources and the vitality of the forest.

Now, in the most basic terms, a forest can be sustained by planting a new sapling for every tree that is removed. However, that is an oversimplified solution.

Proper management of a forest must take into account an assortment of factors, which are assessed by a forest manager, or forester, who is the individual responsible for managing the balance of a forest's environmental, commercial, and recreational viability.

Sustainable forest management addresses forest degradation and deforestation while increasing direct benefits to people and the environment. At the social level, sustainable forest management contributes to livelihoods, income generation and employment. At the environmental level, it contributes to important services such as carbon sequestration and water, soil and biodiversity conservation.

Managing forests sustainably means increasing their benefits, including timber and food, to meet society's needs in a way that conserves and maintains forest ecosystems for the benefit of present and future generations.

Many of the world's forests and woodlands, especially in the tropics and subtropics, are still not managed sustainably. Some countries lack appropriate forest policies, legislation, institutional frameworks and incentives to promote sustainable forest management, while others may have inadequate funding and lack of technical capacity.Where forest management plans exist, they are sometimes limited to ensuring the sustained production of wood, without paying attention to the many other products and services that forests offer.

At the same time, other land uses such as agriculture can seem financially more attractive in the short term than forest management, motivating deforestation and land-use changes.

Examples of Sustainable Forest Management

One of the examples of sustainable forest management that a forest manager might employ to avoid the complete removal of a forest is to use selective logging. Selective logging is the practice of removing certain trees while preserving the balance of the woodland. Selective logging is more time consuming and more expensive then clearing the trees, but it preserves the forest's assets.

Another example of a sustainable forestry practice is allowing young trees time to mature. While a young tree may have value, its value will increase as it matures. Proper forest management will take into account the potential value of trees and delay the harvest of immature trees. In this way, sustainable forest management protects the long-term value of the forest. Other examples of sustainable forestry involve the planting of trees to extend forestlands, as well as the creation of protected forests that provide safe habitats for various plant and animal species. The principles governing sustainable forest management that cover a range of environmental, social, and economic criteria are:

Establish Protected Areas and Conserve Biodiversity

A forest's biodiversity—including its water resources, soils, plant species, and animal populations—must be conserved. This means that forest managers minimize erosion and protect waterways; avoid the use of chemical pesticides; properly dispose of waste; conserve native tree species and maintain genetic diversity on their land; set aside part of their properties as protected areas where logging is prohibited (including forestland that is steeply sloped, provides habitat for critical wildlife species, and serves a culturally or spiritually significant function in the local community); and take other steps to ensure the integrity of the forest.

Prevent Forest Conversion and Protect High Conservation Value Forests

The forest managers should protect natural forests against deforestation, reduce the risk of fires, and take particular care to protect "high conservation value forests." The latter term is used to describe forests that contain significant concentrations of biodiversity; are located in or include rare or endangered ecosystems; are critical providers of ecosystem services; or are fundamental to meeting the basic needs or defining the cultural identity of forest communities.

Have a Management Plan and Harvest Accordingly

Logging activities can take many forms, from selective harvesting to limited, small-scale clear-cutting, which, in temperate forests, can mimic natural disturbances such as fires or landslides. forestry operations must put into place a clearly mapped management plan that specifies the number of trees that can be harvested per acre, and the frequency at which this can occur, based on the growth and regeneration rates of the species found in that ecosystem. The goal is to harvest in such a way that allows these species the chance to regenerate, and ensures that the forest's overall ecological health is maintained, restored, or even enhanced.

Tree Plantations have a Role to Play

Sustainable forestry focuses on keeping natural forests standing, However, the establishment of plantations on already deforested or degraded land can improve the health of an ecosystem and help to meet some of the demand for forest products, taking pressure off of natural forests. plantations must operate according to a management plan that promotes the protection, restoration, and conservation of natural forests.

Use Reduced-impact Logging Techniques

Many people associate logging with the image of a bulldozer leaving behind a denuded landscape, but it is possible to harvest timber without causing collateral damage to other parts of a forest. Reduced-impact techniques allow loggers to fell and extract trees in a manner that reduces damage to other trees in the stand. This approach also minimizes erosion, waste, and carbon emissions.

Train Employees and keep them Healthy

Sawmill training at a certified forestry operation in Cameroon

A forestry business that does not protect its workers is not only unethical, but also unsustainable. Well-trained and healthy employees are essential to ensuring that these enterprises function safely and efficiently. In an examination of community-run forestry businesses in Brazil, certified enterprises did a far better job of protecting their workers than their noncertified peers. Members of certified enterprises were four times more likely to have taken part in a safety course; 94 percent of these businesses offered regular medical exams to their workers; all of the certified enterprises properly washed and stored their protective gear; and 100 percent—four times as many as noncertified enterprises—offered medical attention to their employees when they were injured on the job.

Respect Local Communities and Foster Economic Development

For forestry businesses to be sustainable, they must operate in harmony with their surroundings. This means more than just the natural ecosystems in which they are located; it also applies to the human neighbors with which they co-exist. It means that a certified business must contribute to the social and economic development of a community by offering its members opportunities for employment and compensating indigenous groups for the traditional knowledge that they share regarding forest species and operations. These are not only socially responsible steps, but they also benefit the environment. Providing jobs to local people, for example, can eliminate the incentive to engage in profitable but destructive activities such as wildlife poaching and illegal logging.

Boost Income and Profitability

Sustainable forestry should have a positive economic impact on its practitioners. The steps that help a business earn certification are the same that require the active management of its forestland, teach employees how to work safely and efficiently, and reduce staff turnover, so it's no surprise that these steps can also lead to economic growth.

Coppicing

Regenerating coppiced willow

Coppicing is the process of cutting trees down, allowing the stumps to regenerate for a number of years (usually 7 - 25) and then harvesting the resulting stems.

It makes use of the natural regeneration properties of many tree species, including Oak, Hazel, Maple, Sweet Chestnut, Lime and Ash. Cut such trees down and they will regenerate from the cut stump, producing many new shoots, rather than a single main stem. Regrowth can be exceedingly rapid, with new shoots growing as much as 5cm a day. Oak stems can exceed 2m growth in one season, while Sallow may grow to almost 4m high in the first summer1. The word 'coppice' is derived from the French 'couper' which means 'to cut'. Coppice trees and their produce are known as 'underwood'.

The cut tree stump is known as a stool and the shoots, depending on their harvested size, as rods, poles or logs. The shoots are harvested on a rotational cycle. This means that they are left to grow for a certain number of years and are then cut, whereupon the whole process starts again.

Coppiced woods are either cut according to demand or will more usually be divided into a number of compartments. One of the compartments, or coupes, will be cut each year within the rotational cycle.

The length of the rotational cycle will depend on the tree species, as well as the projected use of the product. A typical rotation might be on a cycle of 7 - 8 years. Many traditional uses require 7 - 15 year old material, although some modern commercial use may take larger material, up to 30 years old.

Coppicing has been practiced in British woodlands for centuries. As a result of the rotational cutting sequence, at any one time there would be coppice at various different stages of regeneration within the woodland. In this way, wood was produced for a variety of uses, in an elegantly sustainable way.

Historically, long rotations were used which would provide really sizeable timber. Coppice products included ships planking, timber for half-timbered Tudor houses, as well as smaller items such as pea and bean sticks, firewood, charcoal, furniture, sheep hurdles, baskets, fencing, hedging sticks, tool handles and brooms.

Coppice management favours a range of wildlife, often of species adapted to open woodland. After cutting, the increased light allows existing woodland-floor vegetation such as bluebell, anemone and primrose to grow vigorously. Often brambles grow around the stools, encouraging insects, or various small mammals that can use the brambles as protection from larger predators. Woodpiles (if left in the coppice) encourage insects such as beetles to come into an area. The open area is then colonised by many animals such as nightingale, European nightjar and fritillary butterflies. As the coup grows, the canopy closes and it becomes unsuitable for these animals again—but in an actively managed coppice there is always another recently cut coup nearby, and the populations therefore move around, following the coppice management.

Overstood sweet chestnut coppice stool, Banstead Woods, Surrey

However, most British coppices have not been managed in this way for many decades. The coppice stems have grown tall (the coppice is said to be *overstood*), forming a heavily shaded woodland of many closely spaced stems with little ground vegetation. The open-woodland animals survive in small numbers along woodland rides or not at all, and many of these once-common species have become rare. Overstood coppice is a habitat of relatively low biodiversity—it does not support the open-woodland species, but neither does it support many of the characteristic species of high forest, because it lacks many high-forest features such as substantial dead-wood, clearings and stems of varied ages. Suitable conservation management of these abandoned coppices may be to restart coppice management, or in some cases it may be more appropriate to use singling and selective clearance to establish a high-forest structure.

Natural Occurrence

Coppice and pollard growth is a response of the tree to damage, and can occur naturally. Trees may be browsed or broken by large herbivorous animals, such as cattle or elephants, felled by beavers or blown over by the wind. Some trees, such as linden, may produce a line of coppice shoots from a fallen trunk, and sometimes these develop into a line of mature trees. For some trees, such as the common beech (*Fagus sylvatica*), coppicing is more or less easy depending on the altitude : it is much more efficient for trees in the montane zone.

Energy Wood

Coppicing of willow, alder and poplar for energy wood has proven commercially successful. The Willow Biomass Project in the United States is an example of this. In this case the coppicing is done in a way that an annual or more likely a tri-annual cut can happen. This seems to maximize the production volume from the stand. Such frequent growth means the soils can be easily depleted and so fertilizers are often required. The stock also becomes exhausted after some years and so will be replaced with new plants. The method of harvesting of energy wood can be mechanized by adaptation of specialized agricultural machinery.

Species and cultivars vary in when they should be cut, regeneration times and other factors. However, full life cycle analysis has shown that poplars have a lower effect in terms of greenhouse gas emissions for energy production than alternatives.

Plants Suitable for Coppicing

Not all trees are plants suitable for coppicing. Generally, broadleaf trees coppice well but most conifers do not. The strongest broad leaves to coppice are:

- Ash
- Hazel
- Oak
- Sweet chestnut
- Lime
- Willow

The weakest are beech, wild cherry and poplar. Oak and lime grow sprouts that reach three feet in their first year, while the best coppicing trees – ash and willow – grow much more. Usually, the coppiced trees grow more the second year, then growth slows dramatically the third.

Coppice products used to include ship planking. The smaller wood pieces were also used for firewood, charcoal, furniture, fencing, tool handles and brooms.

Coppicing Techniques

The procedure for coppicing first requires you to clear out foliage around the base of the stool. The next step in coppicing techniques is to prune away dead or damaged shoots. Then, you work from one side of the stool to the center, cutting the most accessible poles. Make one cut about 2 inches above the point the branch grows out of the stool. Angle the cut 15 to 20 degrees from the horizontal, with the low point facing out from the stool center. Sometimes, you may find it necessary to cut higher first, then trim back.

Forest Gardening

Forest gardening is creating a garden that is deliberately planted to mimic a natural forest ecosystem, except that the species chosen are mainly edible rather than (or as well as) decorative. Some will be chosen for other reasons though – for example firewood, nitrogen fixing or medicines.

Three layers of perennial plants – the herb layer is valerian (medicinal), the shrub layer is flax (fibres), and the canopy is fruit and firewood trees.

So a forest garden is sustainable garden using diverse, perennial edible species, based upon the structure of native woodland, which means that there are layers – from the tops of trees down to the ground, and to the roots under it. Seven layers are generally identified:

- Canopy trees – standard large trees.

- Smaller shade tolerant trees, from dwarf stock, for fruit and nuts.

- Shrubs and bushes such as currants and gooseberries.

- Herbaceous layer of perennial herbs and vegetables.

- Groundcover plants.

- Underground layer – root crops.

- Vertical layer of climbers and vines, beans etc, trained to climb up the trees and bushes.

Harvesting gooseberries from a forest garden, with rhubarb in the foreground and apple and damson trees forming the canopy.

Forest gardening is an ancient practice; there is evidence that people (and animals) have consciously shaped the forests in which they lived for millennia.

Benefits of Forest Gardening

At present most of our food needs are met by giant agri-businesses who use monoculture systems to produce fruit and vegetable crops. These systems are heavily oil and chemical dependent and are slowly eroding and polluting our soils and water courses. Farming may need to change radically quite soon though as oil is used faster than it is discovered.

The benefits of forest garden systems are many:

- Resilient, withstanding drought and flooding through well-developed root and mycorrhizal networks.

- Maintain soil fertility and can be used to reclaim soils that have been polluted.

- Control soil erosion and water runoff.

- Provide their own nutrient requirements, through annual leaf fall, the planting of deep-rooting mineral accumulators (e.g. comfrey) and nitrogen-fixing plants and trees such as Eleagnus, alder and clovers, avoiding the need to constantly import materials, or use chemicals.

- Low maintenance once established.

- The food they provide is nutrient rich and diverse, promoting good health.

- Excellent for wildlife, creating a variety of habitats and attracting beneficial insects.

- Can prevent or remedy soil salinization and acidification.

- Utilize sunlight far more effectively than monoculture systems.

- Attractive, and provide great spaces for play, education and relaxation.

In Tropical Climates

Forest gardens, or home gardens, are common in the tropics, using intercropping to cultivate trees, crops, and livestock on the same land. In Kerala in south India as well as in northeastern India, the home garden is the most common form of land use and is also found in Indonesia. One example combines coconut, black pepper, cocoa and pineapple. These gardens exemplify polyculture, and conserve much crop genetic diversity and heirloom plants that are not found in monocultures. Forest gardens have been loosely compared to the religious concept of the Garden of Eden.

Americas

The BBC's *Unnatural Histories* claimed that the Amazon rainforest, rather than being a pristine wilderness, has been shaped by humans for at least 11,000 years through practices such as forest gardening and *terra preta*. This was also explored in the best-selling book *1491* by author Charles C. Mann. Since the 1970s, numerous geoglyphs have also been discovered on deforested land in the Amazon rainforest, furthering the evidence about Pre-Columbian civilizations.

On the Yucatán Peninsula, much of the Maya food supply was grown in "orchard-gardens", known as pet kot. The system takes its name from the low wall of stones (*pet* meaning circular and *kot* wall of loose stones) that characteristically surrounds the gardens.

The North American ecosystem was managed by the first nations' use of fire to burn underbrush to encourage large game. Large Oak forests harvested for acorns disappeared as the Europeans arrived.

Africa

In many African countries, for example Zambia, Zimbabwe, Ethiopia and Tanzania, gardens are widespread in rural, periurban and urban areas and they play an essential role in establishing food security. Most well known are the Chaga or Chagga gardens on the slopes of Mt. Kilimanjaro in Tanzania. These are an excellent example of an agroforestry system. In many countries, women are the main actors in home gardening and food is mainly produced for subsistence. In North-Africa, oasis layered gardening with palm trees, fruit trees and vegetables is a traditional type of forest garden.

Nepal

In Nepal, the Ghar Bagaincha, literally "home garden", refers to the traditional land-use system around a homestead, where several species of plants are grown and maintained by household members and their products are primarily intended for the family consumption. The term "home garden" is often considered synonymous to the kitchen garden. However, they differ in terms of function, size, diversity, composition and features. In Nepal, 72% of households have home gardens of an area 2–11% of the total land holdings. Because of their small size, the government has never identified home gardens as an important unit of food production, and they thus remain neglected from research and development. However, at the household level the system is very important, as it is an important source of quality food and nutrition for the rural poor and, therefore, is an important contributor to the household food security and livelihoods of farming communities in Nepal. The gardens are typically cultivated with a mixture of annual and perennial plants that can be harvested on a daily or seasonal basis. Biodiversity that has an immediate value is maintained in home gardens as women and children have easy access to preferred food. Home gardens, with their intensive and multiple uses, provide a safety net for households when food is scarce. These gardens are not only important sources of food, fodder, fuel, medicines, spices, herbs, flowers, construction materials and income in many countries, but they are also important for the in situ conservation of a wide range of unique genetic resources for food and agriculture. Many uncultivated, as well as neglected and underutilised species could make an important contribution to the dietary diversity of local communities.

In addition to supplementing diet in times of difficulty, home gardens promote whole-family and whole-community involvement in the process of providing food. Children, the elderly, and those caring for them can participate in this infield agriculture, incorporating it with other household tasks and scheduling. This tradition has existed in many cultures around the world for thousands of years.

Mediterranean Climates

The Mediterranean climate has long, hot, rainless summers and relatively short, cool, rainy winters (Köppen climate classification *Csa*). Its climate conditions are highly variable within an area and modified locally by altitude, latitude, and the proximity to the Mediterranean. In the 1950s the Forest Research Department of the Ministry of Agriculture founded a botanical forest garden in the Sharon region in Israel, the Ilanot

Forest. As the only one of its kind in Israel, it harbours more than 750 species of trees from locations all over the world, including the Japanese sago palm cycas revoluta, fig trees (ficus glomerata), stone pine trees (pinus pinea) that produce tasty pine nuts and adds to the biodiversity of Israel.

Temperate Climates

Robert Hart, forest gardening pioneer

Robert Hart coined the term "forest gardening" during the 1980s. Hart began farming at Wenlock Edge in Shropshire with the intention of providing a healthy and therapeutic environment for himself and his brother Lacon. Starting as relatively conventional smallholders, Hart soon discovered that maintaining large annual vegetable beds, rearing livestock and taking care of an orchard were tasks beyond their strength. However, a small bed of perennial vegetables and herbs he planted was looking after itself with little intervention.

Following Hart's adoption of a raw vegan diet for health and personal reasons, he replaced his farm animals with plants. The three main products from a forest garden are fruit, nuts and green leafy vegetables. He created a model forest garden from a 0.12 acre (500 m²) orchard on his farm and intended naming his gardening method *ecological horticulture* or *ecocultivation*. Hart later dropped these terms once he became aware that *agroforestry* and *forest gardens* were already being used to describe similar systems in other parts of the world. He was inspired by the forest farming methods of Toyohiko Kagawa and James Sholto Douglas, and the productivity of the Keralan home gardens as Hart explains.

From the agroforestry point of view, perhaps the world's most advanced country is the Indian state of Kerala, which boasts no fewer than three and a half million forest gardens. As an example of the extraordinary intensivity of cultivation of some forest gardens, one plot of only 0.12 hectares (0.30 acres) was found by a study group to have twenty-three young coconut palms, twelve cloves, fifty-six bananas, and forty-nine pineapples, with thirty pepper vines trained up its trees. In addition, the small holder grew fodder for his house-cow.

Seven-layer System

The seven layers of the forest garden

Robert Hart pioneered a system based on the observation that the natural forest can be divided into distinct levels. He used intercropping to develop an existing small orchard of apples and pears into an edible polyculture landscape consisting of the following layers:

1. 'Canopy layer' consisting of the original mature fruit trees.

2. 'Low-tree layer' of smaller nut and fruit trees on dwarfing root stocks.

3. 'Shrub layer' of fruit bushes such as currants and berries.

4. 'Herbaceous layer' of perennial vegetables and herbs.

5. 'Rhizosphere' or 'underground' dimension of plants grown for their roots and tubers.

6. 'Ground cover layer' of edible plants that spread horizontally.

7. 'Vertical layer' of vines and climbers.

A key component of the seven-layer system was the plants he selected. Most of the traditional vegetable crops grown today, such as carrots, are sun-loving plants not well selected for the more shady forest garden system. Hart favoured shade-tolerant perennial vegetables.

Further Development

The Agroforestry Research Trust (ART), managed by Martin Crawford, runs experimental forest gardening projects on a number of plots in Devon, United Kingdom. Crawford describes a forest garden as a low-maintenance way of sustainably producing food and other household products.

Ken Fern had the idea that for a successful temperate forest garden a wider range of edible shade tolerant plants would need to be used. To this end, Fern created the organisation Plants for a Future (PFAF) which compiled a plant database suitable for such a system. Fern used the term woodland gardening, rather than forest gardening, in his book Plants for a Future.

The Movement for Compassionate Living (MCL) promote forest gardening and other types of vegan organic gardening to meet society's needs for food and natural resources. Kathleen Jannaway, the founder of MCL, wrote a book outlining a sustainable vegan future called Abundant Living in the Coming Age of the Tree in 1991. The MCL provided a grant of £1,000 to the Bangor Forest Garden project in Gwynedd, North West Wales.

Kevin Bradley coined the phrase "Edible Forest" in the 1980s as the name of his nursery, garden, and orchard on 5 acres in the frigid zone 3 pine forests of northern Wisconsin. Among 3 options, he chose "Edible Forest" because it "evokes at once an ethereal, spiritual, and magical image", of Disney- like "Forest of No Return"; of the biblical "Garden of Eden". This image was perfectly in line with his ongoing experiment begun in 1985 in what he calls a closed loop human environment, combining multi- story tree and field crop "garden/orchards" for maximum beauty and use of space, someday to be very useful in an ever-shrinking world. "The name, at the same time, with its irrational first impression (of course we can't eat a forest), forces the mind to think, if just a little bit, about its inference and thus sticks in our memories". It appeared from Bradley's research that the two words had, prior to the 80's, never been put together before as a noun phrase but which by today, after more than two decades of Bradley's "Edible Forest Nursery" and the 2005 text by Jacke and Toensmeirer's- "Edible Forest Gardens", has grown into a movement and little "Edible Forests" all over the world.

In 2005, Dave Jacke and Eric Toensmeier's two-volume Edible Forest Gardens provided a deeply researched reference focused on North American forest gardening climates, habitats, and species. The book attempts to ground forest gardening deeply in ecological science. The Apios Institute wiki grew out of their work, and seeks to document and share the experience of people around the world working with the species in polycultures.

Permaculture

Bill Mollison, who coined the term permaculture, visited Robert Hart at his forest garden in Wenlock Edge in October 1990. Hart's seven-layer system has since been adopted as a common permaculture design element.

Numerous permaculturalists are proponents of forest gardens, or food forests, such as Graham Bell, Patrick Whitefield, Dave Jacke, Eric Toensmeier and Geoff Lawton. Bell started building his forest garden in 1991 and wrote the book The Permaculture Garden in 1995, Whitefield wrote the book How to Make a Forest Garden in 2002, Jacke and Toensmeier co-authored the two volume book set Edible Forest Gardens in 2005, and Lawton presented the film Establishing a Food Forest in 2008.

Austrian Sepp Holzer practices "Holzer Permaculture" on his Krameterhof farm, at varying altitudes ranging from 1,100 to 1,500 metres above sea level. His designs create micro-climates with rocks, ponds and living wind barriers, enabling the cultivation of a variety of fruit trees, vegetables and flowers in a region that averages 4 °C, and with temperatures as low as -20 °C in the winter.

The primary aims for the forest garden system are:

- To be biologically sustainable, able to cope with disturbances such as climate change.

- To be productive, yielding a number (often large) of different products.

- To require low maintenance.

The crops which are produced will often include fruits, nuts, edible leaves, spices, medicinal plant products, poles, fibres for tying, basketry materials, honey, fuelwood, fodder, mulches, game, sap products. Forest gardens (often called home gardens) have been used for millennia in tropical regions, where they still often form a major part of the food producing systems which people rely on, even if they work elsewhere for much of the time. They may also provide useful sources of extra income. They are usually small in area, often 0.1-1 hectares (0.25-2.5 acres).

In temperate regions, forest gardens are a more recent innovation, over the last 30 years. A major limiting factor for temperate forest gardens in the amount of sunlight available to the lower layers of the garden: in tropical regions, the strong light conditions allow even understorey layers to receive substantial light, whereas in temperate regions this is not usually the case. To compensate for this, understorey layers in temperate forest gardens must be chosen very carefully.

There are plenty of plant crops which tolerate shady conditions, but many are not well known. Many of the more common shrub or perennial crops need bright conditions, and it may be necessary to design in more open clearings or glades for such species. Temperate forest gardens are also usually small in area, from tiny back garden areas up to a hectare (2.5 acres) in size.

The key features which contribute to the stability and self-sustaining nature of this system are:

- The large number of species used, giving great diversity.

- The careful inclusion of plants which increase fertility, such as nitrogen fixers (eg. Alders [Alnus spp], Broom [Cytisus scoparius], Elaeagnus spp, and shrub lupins [Lupinus arboreus]).

- The use of dynamic accumulators – deep rooting plants which can tap mineral sources deep in the subsoil and raise them into the topsoil layer where they become available to other plants, eg. Coltsfoot [Petasites spp], Comfreys [Symphytum spp], Liquorice [Glycyrrhiza spp], Sorrel (and docks!) [Rumex spp].

- The use of plants specially chosen for their ability to attract predators of common pests, eg umbellifers like tansy.

- The use, where possible, of pest and disease resistant varieties, eg. apples.

- The increasing role of tree cover and leaf litter which improve nutrient cycling and drought resistance.

Designing in Layers

A forest garden is organised in up to seven 'layers'. Within these, the positioning of species depends on many variables, including their requirements for shelter, light, moisture, good/bad companions, mineral requirements, pollination, pest-protection, etc. The layers consist of:

Canopy Trees

The highest layer of trees. May include species such as Chestnuts (Castanea spp), Persimmons (Diospyros virginiana), honey locusts (Gleditsia triacanthos), Strawberry trees (Arbutus spp), Siberian pea trees (Caragana arborescens) Cornelian cherries (Cornus mas), Azeroles and other hawthorn family fruits (Crataegus spp), Quinces (Cydonia oblonga), Apples (Malus spp), Medlars (Mespilus germanica), Mulberries (Morus spp), Plums (Prunus domestica), Pears (Pyrus communis), highbush cranberries (Viburnum trilobum).

Small Trees and Large Shrubs

Mostly planted between and below the canopy trees. May includes some of the canopy species on dwarfing rootstocks, and others such as various bamboos, Serviceberries (Amelanchier spp), Plum yews (Cephalotaxus spp), Chinkapins (Castanea pumila),

Elaeagnus spp, and Japanese peppers (Zanthoxylum spp). Others may be trees which will be coppiced to keep them shrubby, like medicinal Eucalyptus spp, and beech (Fagus sylvatica) and limes (Tilia spp) with edible leaves.

Shrubs

Mostly quite shade tolerant. May include common species like currants (Ribes spp) and berries (Rubus spp), plus others like chokeberries (Aronia spp), barberries (Berberis spp), Chinese dogwood (Cornus kousa chinensis), Oregon grapes (Mahonia spp), New Zealand flax (Phormium tenax) and Japanese bitter oranges (Poncirus trifoliata).

Herbaceous Perennials

Several of which are herbs and will also contribute to the ground cover layer by self-seeding or spreading. These may include Bellflowers with edible leaves (Campanula spp), Comfreys (Symphytum spp), Balm (Melissa officinalis), Mints (Mentha spp), Sage (Salvia officinalis), and Tansy (Tanacetum vulgare).

Ground Covers

Mostly creeping carpeting plants which will form a living mulch for the 'forest floor'. Some may be herbaceous perennials, others include wild gingers (Asarum spp), cornels (Cornus canadensis), Gaultheria spp, and carpeting brambles (eg. Rubus calycinoides & R.tricolor).

Climbers and Vines

These are generally late additions to the garden, since they obviously need sturdy trees to climb up. They may include hardy kiwis (Actinidia spp), and grapes (Vitis spp).

Rhizosphere

Any design should take account of different rooting habits and requirements of different species, even if root crops are not grown much. Some perennials with useful roots include liquorice (Glycyrrhiza spp) and the barberries (Berberis spp) whose roots furnish a good dye and medicinal products. Various beneficial fungi can also be introduced into this layer.

Pollarding

Pollarding is a woodland management method of encouraging lateral branches by cutting off a tree stem or minor branches two or three metres above ground level.

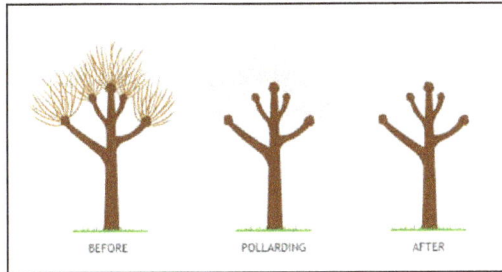

The tree is then allowed to regrow after the initial cutting, but once begun, pollarding requires regular maintenance by pruning. This will eventually result in a somewhat expanded (or swollen) top to the tree trunk with multiple new side and top shoots growing from it.

A tree that has been pollarded is known as a pollard. A tree which has been allowed to grow without being cut as a pollard (or coppice stool) is called a maiden or maiden tree. Pollarding older trees may result in the death of the tree, especially if there are no branches below the cut, or the tree is of an inappropriate species. Pollarding is sometimes abused in attempts to curb the growth of older or taller trees but when performed properly it is useful in the practice of arboriculture for tree management.

Pollard trees may attain much greater ages than maiden trees because they are maintained in a partially juvenile state, and they do not have the weight and windage of the top part of the tree. Older pollards often become hollow, and so can be difficult to age accurately. Pollards tend to grow more slowly than maiden trees, with narrower growth rings in the years immediately after cutting.

Pollarding is a pruning technique used for many reasons, including:

- Preventing trees and shrubs outgrowing their allotted space.

- Pollarding can reduce the shade cast by a tree.

- May be necessary on street trees to prevent electric wires and streetlights being obstructed.

These are a few of the plants it can be used on:

- Ash (Fraxinus)

- Common lime (Tilia × europaea)

- Elm (Ulmus)

- Elder (Sambucus)

- Gum (Eucalyptus)

- London plane (Platanus × hispanica)

- Mulberry (Morus)

- Oak (Quercus)

- Some species of Acer (A. negundo and its cultivars)

- Tulip tree (Liriodendron).

Pollarding a tree is usually done annually, and would need to be carried out every few years to avoid potential problems. This usually involves hiring an arborist, so can be expensive. Why not consider the following before pollarding:

- Plant a tree small enough to fit its allotted space. This will only need minimal pruning.

- Try other pruning options suitable for large trees, such as crown thinning or crown reduction.

When to Pollard

The best time for pollarding many trees and shrubs is in late winter or early spring. However, bear in mind the following:

- Avoid pruning Acer species in spring when they are prone to bleeding sap. Summer can be a suitable time to pollard. However, the new growth may be poor as a result of the scorch, drought or heavy shade cast by neighbouring trees.

- The least favourable time for pollarding is the autumn, as decay fungi may enter the pruning cuts.

Pollarding Method

Young Trees

Once young trees or shrubs have reached the desired height, you can begin to pollard them. This involves choosing a framework:

- On a shrub, this might be one stem cut to a metre high – a mass of stems will grow from the top.

- With a tree, it is more typical to leave a trunk supporting three or five branches – these branches are cut back to a desirable length and the twiggy growth appears at these ends.

Initially, the new branches are held weakly in place as they grow rapidly from underneath the bark, rather than from within the tree. As the wood lays down annual growth rings, the union strengthens, often forming a thickened base where the shoot meets the trunk. Over a number of years, a swollen 'pollard head' forms where new shoots grow each year.

Maintaining a Pollard

Once a tree or shrub is pollarded, continue the annual cycle of cutting.

- Branches should be pruned just above the previous pollarding cuts.

- In some cases, such as where some leaf cover is required, leave some branches intact or cut back to a side branch.

Rejuvenating an Overgrown Pollarded Tree

Seek advice from an arborist before doing any work. Although having a tree pollarded regularly is expensive, an overgrown pollard may require more surgery to remove larger parts of the tree at a greater height.

Try the following to rejuvenate an overgrown pollarded tree or shrub:

- Remove any spindly and weakly-attached branches.

- Consider whether the branches can be thinned out, and reduced in length, to create a tree-like framework, effectively restoring the pollard to a tree.

- It may be possible to remove all the branches that have grown from the stumps of the old pollards. London plane (Platanus × hispanica) responds to this treatment.

- Horse chestnut (Aesculus × hippocastanum) needs to be cut to a higher point in the tree, rather than to the original pollards. This avoids exposing large amounts of old wood, but creates a second set of pollard heads.

- In some cases, such as with hornbeam (Carpinus betulus) and ash (Fraxinus excelsior), it is beneficial to retain some of the branches. Likewise, some oak trees, such as Quercus robur and Q. petraea, do best when substantial portions of their main branches are retained.

After any major work, the tree should be monitored for any further maintenance required.

Short Rotation Coppice

Short Rotation Coppice (SRC) is a farming method to cultivate fast growing trees. The main characteristic of SCR species is their ability to sprout again from their roots after harvesting. These plantations can be used to clean pre-treated domestic wastewater: the biologic activity in the soil purifies the wastewater and the plants can absorb nutrients. The wastewater can be spread on the fields using conventional irrigation systems. This can increase the yield of the trees up to 100%.

Advantages

- Provides a second source of income to farmers.
- It is a low technology system.
- Supporting sustainable rural development.
- Production of renewable biomass as a fuel.

Disadvantages

- Bridge strong frost periods.
- Large area required, implementation only in rural areas.
- The particle size are limited.
- Only for domestic wastewater.

Design and Construction Principles

For these plantations species like willows, poplars, eucalyptus or bamboo are used, since they are fast growing in their youth and can sprout again from their roots after harvesting. The plantations are arranged in a single or double row system. An average yield of 10 tons absolute dry wood per year and hectare can be expected. This has the energy content equivalent to 5,000 L heating oil. The harvesting is usually conducted in intervals between 3 to 5 years depending on the formation and the tree species. Since water is their main growth limitation factor, on nutrient poor and dry soil yield can be increased significantly by adding nutrient-rich wastewater. The chipped harvested wood is an excellent fuel, which can be used in regional power plants, district heating systems or households. While constructed wetlands focus mainly on wastewater treatment and are sealed at their base for groundwater protection, the advantage of SRCs over constructed wetlands lies in the combined wastewater treatment and the production of wooden biomass, which means an additional income for farmers. A SRC represents an open-bottom fixed-bed reactor of a construction height of between 1.0 and 1.5 m resulting in an effective reduction of pathogens. To avoid a nutrient overload it is important to control and document the amount and quality of the applied wastewater and sewage sludge.

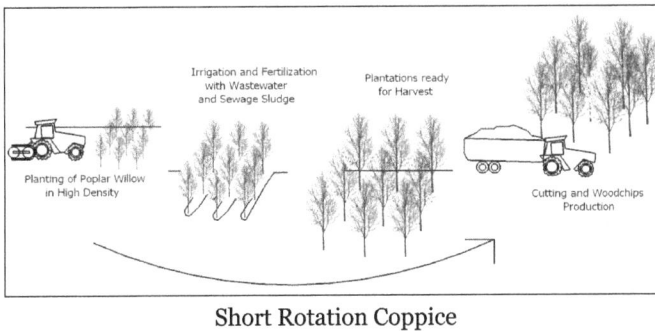

Short Rotation Coppice

Operation and Maintenance

For this treatment system the main requirement is land. In Europe, cost-effectiveness using fully mechanical planting and harvesting systems is reached starting at 5 ha. The plantation of a SRC and its harvest can be also done manually. Cuttings from tree nurseries are required as seedlings. In one hectare up to 12,000 trees like willows, popular or eucalyptus can be grown. For the distribution of the wastewater a drip irrigation system and slurry pump are needed. Before the wastewater enters the system, a mechanical pre-treatment would be needed to filter and avoid clogging. To maintain SRC, typical farmer skills like knowledge about plants, fertilizer, irrigation, and familiarity with agricultural machinery are needed.

A number of different species are suitable for short rotation coppicing, with different optimum cycle periods.

SRC Willow

Varieties

Willow (Salix spp.) is planted as rods or cuttings in spring using specialist equipment at a density of 15,000 per hectare. The willow stools readily develop multiple shoots when coppiced and several varieties have been specifically bred with characteristics well suited for use as energy crops. Information and advice is available from the Forest Research Yield models for energy coppice of poplar and willow website.

Growth

During the first year it can grow up to 4m in height, and is then cut back to ground level in its first winter to encourage it to grow multiple stems.

The first harvest is in winter, typically three years after cut back, again using specialist equipment, however a cycle of 2 or 4 to 5 years is also common.

In fertile sites growth can be very strong during the first two years after coppicing, giving rapid site capture, reducing thereafter and so a 2 year cutting cycle may be more appropriate.

Yield

Yield is dependent on many factors, including:

- Site
- Water availability
- Weed control
- Planting density
- Light
- Temperature.

Harvesting

Typically the first harvest may be expected to be somewhat lower than subsequent ones, and figures from 7 to 12 oven dried tonnes per hectare per annum can be expected on reasonably good sites.

Harvesting may be as rods (up to 8 m length), billets (5-15 cm lengths) or as direct chip harvesting. Direct chip harvesting can cause problems for storage with rapid composting (and hence loss of energy content) and mould formation (and attendant health risks) owing to the high moisture content of freshly harvested willow. This can be less of a problem with billets owing to improved air flow through the pile.

Viability

A willow SRC plantation may be expected to be viable for up to 30 years before it becomes necessary to replant and can reach 7-8 m in height at harvest. The site should be reasonably flat, or with a slope no more than 7% and, to be eligible for the Defra Energy Crops Scheme (ECS) grant, needs to be at least 3 ha, though this need not all be in a single plot.

SRC Poplar

Growth

Poplar (Populus spp.) displays more apical dominance than willow and is therefore less ready to develop multiple stems following coppicing. Shoots can reach up to 8 m by the end of the first rotation. It therefore tends to develop fewer, thicker stems than willow, and consequently has a lower bark to wood ratio. Individual shoots can reach up to 8 m by the end of the first 3 year rotation.

Planting

Poplar is planted in spring, from cuttings. These cuttings must have anapical bud within 1 cm of the top of the cutting. Because of this it is difficult to use poplar in equipment developed for plantingwillow short rotaion coppice.

Planting density is lower than for willow, typically 10-12,000 per ha. Cut back takes place late in the following winter.

Yield

Yield is very site dependent, and in some sites can out perform willow. Average yield on a suitable site is likely to be in the region of8 oven dry tonnes per hectare per year.

Harvesting

Poplar responds well to harvesting cycles of around four or five years which is slightly longer than the 3 years often recommended for willow. This is because growth in the first year following cutback or harvest is generally not as rapid as in subsequent years. Combined with a very up right growth habit this means that the crop may not develop a closed canopy, and hence maximum light interception, until the second or third year.

Harvesting requires similar equipment to willow, however, owing to the tendency of poplar to form fewer, heavier stems, it must be slightly more robust.

Removal of a poplar crop at the end of the useful life of the plantation can bemore difficult than for willow as poplar often forms a large taproot which will generally require a large excavator to remove or more time to decay naturally.

Broadleaf Coppice

Many traditional broadleaf species can be grown as coppice, such as ash, hazel, sweet chestnut, sycamoreetc. The rotation is typically longer than for willow or poplar, and the yield is lower, but for management within a mixed woodland this can both help to produce products on a more regular basis than from conventional forestry, and also may be useful to produce smaller diameter logs and stems for both firewood and traditional coppice markets.

References

- What-is-sustainable-forest-management-definition-and-examples: study.com, Retrieved 2 April, 2019

- Forestry: fao.org, Retrieved 22 June, 2019

- What-is-sustainable-forestry: rainforest-alliance.org, Retrieved 8 August, 2019

- Coppice, woodland-manage: countrysideinfo.co.uk, Retrieved 17 January, 2019

- What-is-coppicing, trees: gardeningknowhow.com, Retrieved 15 March, 2019

- Forest-gardening: lowimpact.org, Retrieved 25 May, 2019

- Forest-gardening: agroforestry.co.uk, Retrieved 23 February, 2019

- Rackham, Oliver (2003). Ancient Woodland; its history, vegetation and uses in England (New Edition). Castlepoint Press. ISBN 1-897604-27-0

- Pollarding, tree-pruning: heritagearboriculture.co.uk, Retrieved 11 July, 2019

- Short-rotation-coppice, appropriate-technologies: sswm.info, Retrieved 8 January, 2019

- Short-rotation-coppice, energy-crops, fuel, biomass-energy-resources: forestresearch.gov.uk, Retrieved 2 March, 2019

Permissions

Index

www.ingramcontent.com/pod-product-compliance
Lightning Source LLC
Chambersburg PA
CBHW061957190326
41458CB00009B/2893